Probiotics 2

Probiotics 2

Applications and practical aspects

Edited by

R. Fuller

Freelance Consultant in Gut Microecology
Reading, UK

CHAPMAN & HALL

London · Weinheim · New York · Tokyo · Melbourne · Madras

Published by Chapman & Hall, 2–6 Boundary Row, London SE1 8HN

Chapman & Hall, 2–6 Boundary Row, London SE1 8HN, UK

Chapman & Hall GmbH, Pappelallee 3, 69469 Weinheim, Germany

Chapman & Hall USA, 115 Fifth Avenue, New York, NY 10003, USA

Chapman & Hall Japan, ITP-Japan, Kyowa Building, 3F, 2-2-1 Hirakawacho, Chiyoda-ku, Tokyo 102, Japan

Chapman & Hall Australia, 102 Dodds Street, South Melbourne, Victoria 3205, Australia

Chapman & Hall India, R. Seshadri, 32 Second Main Road, CIT East, Madras 600 035, India

First edition 1997

© 1997 Chapman & Hall

Typeset in 10/12pt Palatino by Acorn Bookwork, Salisbury, Wiltshire
Printed in Great Britain by T. J. International Ltd.

ISBN 0 412 73610 1

∞ Printed on acid-free text paper, manufactured in accordance with ANSI/NISO Z39.48-1992 (Permanence of Paper).

Contents

Contributors

R. Fuller, Russet House, 59 Ryeish Green, Three Mile Cross, Reading, RG7 1ES, UK

G.R Gibson, Institute of Food Research, Early Gate, Whiteknights Road, Reading, RG6 2EF, UK

J.M. Saavedra, Johns Hopkins Hospital, 6000 North Wolfe Street, Brady 320, Baltimore MD 2128-2631, USA

S. Macfarlane and G.T. Macfarlane, Medical Research Council, Dunn Clinical Nutrition Centre, Cambridge, UK

G. Corthier, Institut National de la Recherche Agronomique, Unité d'Ecologie et de Physiologie du System Digestif, 78352 Jouy-en-Josas Cedex, France

P. Marteau, T. Vesa and J.-C. Rambaud, Hôpital Saint-Lazare, 107 bis, rue du fg Saint-Denis, 75475 Paris Cedex, France

A. Hosono, Shinshu University, Minamiminouwa, Nagano, 399-45, Japan

H. Kitazawa and T. Yamaguchi, Department of Animal Science, Faculty of Agriculture, Tohoku University, Sendai-Aobaku, 981, Japan

G. Famularo, S. Moretti, S. Marcellini and C. De Simone, Università de L'Aquila, Cathedra di Malattie Infettive, 671000 L'Aquila, Italy

J.T. Huber, Department of Animal Sciences, University of Arizona, 205 Shantz Building, Tucson, Arizona 85721, USA

R.W.A.W. Mulder, DLO Institute of Animal Sciences and Health, Wageningen, The Netherlands

R. Havenaar, TNO Nutrition and Food Research Institute, Zeist, The Netherlands

J.H.J. Huis in 't Veld, Department of the Science of Food of Animal Origin, University of Utrecht, The Netherlands

CHAPTER 1

Introduction

R. Fuller

1.1 DEVELOPMENT OF COMMERCIAL PREPARATIONS

The history of the probiotic effect has been well documented many times previously (see e.g. Bibel, 1982; Fuller, 1992). The consumption of fermented milks dates from pre-biblical times but the probiotic concept was born at the end of the last century with the work of Metchnikoff at the Pasteur Institute in Paris.

In the century that has elapsed since Metchnikoff's work, the probiotic concept has been accepted by scientists and consumers throughout the world. Attempts to refine the practice from the use of traditional soured milks to preparations containing specific micro-organisms have occupied the thoughts and endeavours of scientists in many different countries. But, in spite of the large amount of effort expended in attempting to explain and define the effect, it has to be admitted that little is known of the way in which probiotics operate. There are likely to be several different mechanisms because it seems highly improbable that a mode of action that explains resistance to microbial infection will also hold true for improved milk production or alleviation of lactose malabsorption.

The lack of fundamental knowledge about the mechanism of the probiotic effect has not deterred the development of a great many probiotic preparations destined for treatment of various conditions in man and animals. There are currently over 20 products on the market in the UK. The dearth of basic information about the probiotic effect has meant that much of the development has been empirical and not always based on sound scientific principles. One factor that has been used in the selection of probiotic cultures has been the ability to adhere to gut epithelial cells of the animal to which the probiotic is being fed.

Probiotics 2: Applications and practical aspects.
Edited by R. Fuller.
Published in 1997 by Chapman & Hall. ISBN 0 412 73610 1

There was a good correlation found between *in vitro* and *in vivo* results for adhesion of different strains of bifidobacteria (Crociani *et al.*, 1995). Adhesion is now generally accepted as an important colonization factor and since establishment in the intestine is an essential prerequisite of effective probiotic activity, it is to be applauded as the first step in a rational approach to the selection of micro-organisms for inclusion in probiotic preparations.

However, it should be appreciated that attachment is not an essential attribute of a successful probiotic organism. Rapid growth can achieve the same end. In some cases such as fungi (*Saccharomyces cerevisiae* and *Aspergillus oryzae*) the effect is gained without either attachment or rapid growth; mere survival is adequate. Under these conditions continuous administration is required for the maximum realization of the probiotic effect.

The numerous probiotic products on the market claim to have many different effects including improved resistance to infectious disease, antitumour activity, increased growth rate and feed conversion in farm animals, improved milk production by cows and increased egg production by poultry. The range of micro-organisms contained in the probiotic preparations is wide, comprising bacteria, moulds and yeasts.

By far the most frequently used of these three groups are the bacteria with lactic acid bacteria (lactobacilli, streptococci, enterococci and bifidobacteria) the predominant genera. The emphasis on the lactic acid bacteria stems partly from the fact that there is evidence that the lactic acid bacteria occupy a central role in the gut flora, which enables them to influence the composition of the flora to the benefit of the host. This has been elegantly demonstrated by the work of Tannock and his group in New Zealand. By developing a population of mice with a lactobacillus-free gut microflora, they have been able to study the way in which the gut lactobacilli can affect the metabolism and growth of the host animal which, in this case, is the mouse. Their results are summarized in Table 1.1.

The use of lactic acid bacteria was also stimulated by the early work of Metchnikoff who espoused the view that the normal gut microflora

Table 1.1   Effect of lactobacilli on mice (from Tannock, 1995)

Increased bile salt hydrolase activity
More unconjugated bile acids
Growth rate unaffected
Azoreductase activity reduced
β-Glucuronidase activity reduced
Enzyme activities associated with duodenal enterocytes unaffected
Serum cholesterol concentration unaffected

was having an adverse effect on the host and that these harmful effects could be ameliorated by consuming soured milks. Although the word was not coined until many years later, these could claim to be the first probiotics. In the form of yoghurt and more recently, bio-yoghurt, the fermented milk preparations persist to the present day although the micro-organisms used for the fermentation of the milk have changed.

Probiotic preparations vary in the way in which they are presented; they may be in the form of powders, tablets, pastes or sprays with different excipients to maintain the preparation in the required condition. The type of preparation employed is determined by the way in which it is intended to use the probiotic. For example, pastes are used for individual dosing of calves and pigs, whereas sprays may be used to treat day-old chicks *en masse* in boxes. The microbial content of the preparations varies, with some containing only one organism while others have up to seven different species. It is, therefore, often not possible to compare the effects obtained using one probiotic with those given by another. A panel of experts convened to discuss the effect of consuming lactic cultures on health concluded that 'the optimal prophy-lactic culture may be mixed: different strains can be targeted toward different ailments and can be blended into one preparation' (Sanders, 1993a).

1.2 FACTORS AFFECTING THE RESPONSE

The results of trials set up to substantiate the claims made for probiotics are often inconsistent. Reviewing the effect of probiotics on diarrhoea, Sanders (1993b) concluded that of 14 studies which she had analysed only three gave a definite positive result. Of the remainder, five were negative and six more were positive but, because of the poor experi-mental design and data analysis, they were of questionable significance. Variations may occur, even between trials of the same product. These results are not as incompatible as they at first appear. They may be related to some change in composition of the preparation being used (e.g. loss of viability) or in its method of administration which makes direct comparison between the two trials impossible.

It is important to be aware of the various factors that can change the response to a probiotic (Table 1.2). With this sort of information in mind, the results detailed in the following chapters of this book can be more satisfactorily assessed and the significance of any inconsistencies better appreciated. The way in which a probiotic is prepared can affect its subsequent performance when used in the field. The growth condi-tions and the point of harvest can both influence survival and behaviour of the product. Similarly, the pressures and temperatures

Table 1.2   Causes of variation in probiotic trials

Method of preparation
Storage conditions
Contamination
Poor viability
Incorrect taxonomy
Status of the gut microflora
Frequency of dosing
Growth phase of the host animal
Survival in the intestinal tract
Changes in diet
Degree of stress

applied during pelleting or tableting can be important. These factors have been expertly discussed by Lauland (1994).

Subsequently, the conditions of storage will affect the viability and the shelf-life of the product. Absence of moisture is of paramount important if the product is to maintain its viable count, and storage under vacuum or nitrogen is recommended. Unfortunately, the precautions needed to maintain viability are not always observed and the number of viable cells in some products is below the level claimed.

In a recent publication (Hamilton-Miller, Shap and Smith 1996), only two out of the 13 probiotic preparations studied matched the specifications on the label. These were both single-strain preparations containing *Lactobacillus acidophilus*. Multistrain preparations can present difficulties of enumeration of subdominant species that have been suppressed on the culture plate by dominant groups. The maintenance of viability over time giving an acceptable shelf-life is an absolutely fundamental property of a successful probiotic.

It is also obviously important that the product contains only the organisms that are listed in the product description, but some products fall short in this respect. Not only are there micro-organisms present that should not be there (i.e. contaminants), but the active organisms that should be there are sometimes completely absent. In a study of probiotics recommended for recolonizing the vagina, it was found that of the 13 preparations that listed *L. acidophilus* as the dominant microbe, only five were found to contain it (Hughes and Hillier, 1990). Of the 16 preparations tested, 11 harboured organisms that were not supposed to be there.

Modern developments in microbial taxonomy can also cause confusion. For example, what was once known as *Lactobacillus bulgaricus* is now *L. delbrueckii* subsp. *bulgaricus*, *Streptococcus faecium* is now *Enterococcus faecium* and many strains of *L. fermentum* have been reclassified as *L. reuteri*. This sort of scientifically based updating of nomenclature can make back-reference difficult. Nor should taxonomic similarity be assumed to confer identical metabolic properties on two strains of the same species. For example, when three different strains of *L. acidophilus* were tested for alleviation of lactose maldigestion, only one was found to be effective (Lin, Savaiano and Harlander, 1991) as measured by reduction of hydrogen in the breath. Two different strains of the same species can differ in amount of lactic acid produced or the ability to adhere to gut epithelial cells. It is, therefore, not surprising that two probiotic preparations based on the same species can give different results.

Diet can influence the way in which a micro-organism affects the host. For example, enteric bacteria which, on their own, have no adverse effect on gnotobiotic rats reduced their growth when kidney bean lectin was administered (Pusztai *et al.*, 1990). This sort of interaction could lead to a situation where the probiotic was only effective when certain dietary components are present.

The features discussed so far relate to the micro-organisms in the preparation, but there are also host factors which can influence the outcome of the probiotic trial.

The microbiological status of the animal can vary and if the organisms responsible for the condition (e.g. growth depression) that is to be reversed by the probiotic are absent, then there is no potential for improvement and the probiotic will appear to be inactive. A corollary of this is that probiotics will show less or no effect under good management conditions that entail a high level of hygiene. This is an effect that can also be demonstrated for growth promotion of chickens by dietary supplements of antibiotics.

The age, or growth phase, of the animal is also important. Different responses to the same probiotic have been recorded in day-old and adult chickens, suckling and weaned pigs and lactating and non-lactating cows. Some of this type of variation may be accounted for by differences in the gut microflora which is known to change as the animal gets older and can also be affected by dietary changes. For example, the suckling and weaned pigs would be consuming very different diets. The trend towards early weaning means that at weaning the piglet is deprived of the passive protection afforded by the antibodies in the sow's milk at a time when it is not fully immune-competent. However, some of the variation may also reflect changes in the basic physiology and metabolism of the animal which may affect the ability of the probiotic organisms to survive in the intestinal tract. Stress may also have a similar effect by inducing changes in the gut habitat mediated by hormones that reduce the mucous lining of the gut.

The method of dosing must also be taken into consideration when assessing the results of probiotic trials. The right choice must be made between using powder incorporated in the diet or water; pastes or tablets for individual dosing; or sprays for incidental acquisition from the environment. If an inappropriate preparation is used the result will be affected. The frequency of dosing can also vary and this can influence the outcome of a trial. At present, there is no sound evidence to indicate what is the minimum amount of probiotic required to give the full probiotic response. One of the few pieces of work done to explore the dose response of a probiotic showed that the faecal count of *Lactobacillus* GG was improved when the administered dose was 1.2×10^{10} c.f.u. per day compared with 2.1×10^9 c.f.u. (Saxelin, Ahokas and Salminen, 1993). This result was obtained with humans and cannot be directly transposed to any other animal species or any other probiotic.

It is a generally held principle that colonization of the gut will be more readily attained if the organism being administered originates from the gut and, more particularly, if it originates from the same species of animal as that being dosed (Fuller, 1978). This so-called host-specific effect has been made much of in the past but recently non-gut organisms have been used effectively. Indeed, Metchnikoff's original observations were obtained using dairy micro-organisms (such as what is now called *L. delbrueckii* subsp. *bulgaricus*). In later years there was a move to *L. acidophilus* on the assumption that, being of gut origin, it would colonize the gut more effectively. However, *Saccharomyces cerevisiae* and *Aspergillus oryzae* are now being used very successfully as probiotics for cattle. Although they do not colonize the rumen, they survive and have significant effects on rumen metabolism which, in turn, affects milk production. Similarly, *Sac. boulardii* is a very effective treatment for antibiotic-associated diarrhoea (McFarland *et al.*, 1995). In a double-blinded, placebo-controlled study, treatment with *Sac. boulardii* gave a significant decrease in the incidence of diarrhoea.

The composition of the gut microflora is in a very dynamic state with the dominant strain composing the total count of a particular species changing periodically. For example, in the piglet the lactobacillus count may remain constant but by using plasmid-profiling techniques a succession of different types was detected (Tannock, Fuller and Pederson, 1990). Naiti *et al.* (1995) also found different lactobacilli predominating during the first 4 weeks of life in the pig. *L. reuteri* appeared on the first day after birth and *L. acidophilus* was not detectable until the pigs were 7 days old. Such an unstable condition would seem to argue against the importance of using host-specific strains because the survival in the gut is limited by endogenous factors.

However, the choice of a good colonizer can still affect the outcome, and attention to colonization factors such as epithelial attachment and resistance to acid will optimize the survival of the administered organisms in the gut. Thus, while it may still be necessary to feed the probiotic continuously, the effectiveness of each dose will be maximized if one uses an organism which can resist the various antimicrobial agents present in the gut. Obviously, the use of an organism normally found in symbiotic association with the host will better ensure its survival and persistence in the gut. Recent work by Tannock (pers. comm.) using ribotyping techniques has identified strains in the human gut that persisted for 1 year. So it may be possible by careful selection to choose strains that will maintain high counts in the gut for long periods. But, in most instances, it seems that however carefully the probiotic strain is selected, it is very unlikely that it will be able to colonize the gut permanently.

1.3 FUTURE DEVELOPMENTS

The maintenance of viability both on the shelf and in the gut is a continuing aim for probiotic manufacturers. One development that would obviate this difficulty would be the discovery of the mode of action of probiotics. This might make it possible to use the chemical agent responsible for the probiotic effect to replace the live organisms now being used which produce the active agents under *in vivo* conditions. This will not be effective if the agent is susceptible to digestion by gut or bacterial enzymes before it reaches the target site in the gut. In this respect the *in situ* production of the probiotic agent may never be replaced by a chemical agent. However, recent developments have suggested that at least in some aspects the new approach might work. A new generation of microbial stimulants has been produced that have a specific stimulatory effect on the bifidobacteria of the lower gut. These so-called *prebiotics* (Gibson and Roberfroid, 1994) are oligosaccharides produced from various substrates.

Useful supplements might also be generated by work on the identification of the factor produced by lactobacilli which prevents adhesion of *Escherichia coli* to epithelial cells. Recent work on *L. fermentum* identified a carbohydrate which inhibited adhesion of *E. coli* K88 to porcine ileal mucus (Couwehand and Conway, 1996). Although this substance is unlikely to be produced by lactobacilli under *in vivo* conditions, it is an approach that might yield an effective new type of prebiotic.

The attraction of this sort of supplement is that it would remove the necessity for producers to maintain viability over long periods and would allow the industry to produce the sterile products with which they are so much more familiar. Thus, one of the causes of variability (variation in viable count) would be removed. The absence of viable organisms in the product would also allow genetic manipulation techniques to be applied to increase production of the active substance, without the attendant problems associated with release of genetically modified organisms into the environment.

The development of sterile preparations might also increase the potential of the effects which probiotics have on immunity. Recent work has shown that orally administered lactobacilli can improve immune status by increasing circulating and local antibody levels, gamma interferon concentration, macrophage activity and the numbers of natural killer cells. Findings such as these broaden the scope for the effects that probiotics can have on the host animal. The benefits are no longer restricted to the gastrointestinal tract but have the potential to protect against disease and other adverse effects occurring in other parts of the body. Future efforts might be directed towards determining how to maximize the immunostimulatory effect of probiotic preparations. For example, how important is epithelial attachment and translocation to the realization of the full effect on the immune system? The development of an active sterile preparation might also enable the *in vivo* concentration to be increased above that which is attainable by the present *in vivo* production. Depending on its metabolic side-effects, the sterile agent could be administered parenterally, giving it improved access to the immune system.

1.4 CONCLUSIONS

The basis of the probiotic effect is scientifically sound but it is essential, if the full potential of the practice is to be realized, that they are used in the correct way. Attention must be paid to the factors discussed above that can affect the result obtained in an animal trial or clinical test. If this is done, probiotics can work and have important beneficial effects in animals and man. There are bound to be inconsistencies, but the whole concept should not be discarded merely because some trials have failed to give the desired result. There is, if sufficient thought is given to the problem, frequently a rational explanation for the different results obtained in two trials. We should look very carefully at the conditions prevailing in successful trials and reproduce them when subsequently testing probiotics.

1.5 REFERENCES

Bibel, D.J. (1982) Bacterial interference, bacteriotherapy and bacterioprophylaxis, in *Bacterial Interference* (ed. R. Aly and H.R. Shinefield), CRC Press, Florida, pp. 1–12.

Couwehand, A.C. and Conway, P.L. (1996) Purification and characterisation of a component produced by *Lactobacillus fermentum* that inhibits the adhesion of K88 expressing *Escherichia coli* to porcine ileal mucus. *J. Appl. Bacteriol.*, **80**, 311–18.

Crociani, J., Grill, J.P., Huppert, M. and Ballongue, J. (1995) Adhesion of different bifidobacteria strains to human enterocyte-like Caco-2 cells and comparison with *in vivo* study. *Letters Appl. Microbiol.*, **21**, 146–8.

Fuller, R. (1978) Epithelial attachment and other factors controlling the colonisation of the intestine of the gnotobiotic chicken by lactobacilli. *J. Appl. Bacteriol.*, **45**, 389–95.

Fuller, R. (1992) History and development of probiotics, in *Probiotics: The Scientific Basis* (ed. R. Fuller), Chapman & Hall, London, pp. 1–8.

Gibson, G.R. and Roberfroid, M.B. (1994) Dietary modulation of the human colonic microbiota: introducing the concept of prebiotics. *J. Nutr.*, **125**, 1401–12.

Hamilton-Miller, J.M.T., Shap, P. and Smith, C.T. (1996) 'Probiotic' remedies are not what they seem. *Brit. Med. J.*, **312**, 55–6.

Hughes, V.L. and Hillier, S.L. (1990) Microbiologic characteristics of *Lactobacillus* products used for colonisation of the vagina. *Obst. Gynaecol.*, **75**, 244–8.

Lauland, S. (1994) Commercial aspects of formulations, production and marketing of probiotic products, in *Human Health: The Contribution of Microorganisms* (ed. A.W. Gibson), Springer, London, pp. 159–73.

Lin, M.Y., Savaiano, D. and Harlander, S. (1991) Influence of non-fermented dairy products containing bacterial starter cultures on lactose maldigestion in humans. *Dairy Sci.*, **74**, 87–95.

McFarland, L.V., Surawicz, C.M., Greenberg, R.N. *et al.* (1995) Prevention of β-lactam-associated diarrhoea by *Saccharomyces boulardii* compared with placebo. *Am. J. Gastroenterol.*, **90**, 439–48.

Naiti, S., Hayadashidani, H., Kaneko, K. *et al.* (1995) Development of intestinal lactobacilli in piglets. *J. Appl. Bacteriol.*, **79**, 230–6.

Pusztai, A., Grant, G., King, T.P. and Clark, E.M.W. (1990) Chemical probiosis, in *Recent Advances in Animal Nutrition* (ed. W. Haresign and D.J.A. Cole), Butterworth, London, pp. 47–60.

Sanders, M.E. (1993a) Summary of conclusions from a consensus panel of experts on healthy attributes of lactic cultures: significance of fluid milk products containing cultures. *J. Dairy Sci.*, **76**, 1819–28.

Sanders, M.E. (1993b) Effect of consumption of lactic cultures on human health. *Adv. Fd. Nutr. Res.*, **37**, 67–130.

Saxelin, M., Ahokas, M. and Salminen, S. (1993) Dose response on the faecal colonisation of *Lactobacillus* strain GG administered in two different formulations. *Micro. Ecol. Hlth. Dis.*, **6**, 119–22.

Tannock, G.W. (1995) Microecology of the gastrointestinal tract in relation to lactic acid bacteria. *Intern. Diary J.*, **4**, 1059–70.

Tannock, G.W., Fuller, R. and Pederson, K. (1990) Lactobacillus succession in the piglet's digestive tract demonstrated by plasmid profiling. *App. Environ. Microbiol.*, **56**, 1310–16.

CHAPTER 2

Probiotics and intestinal infections

G.R. Gibson, J.M. Saavedra, S.Macfarlane and G.T. Macfarlane

2.1 INTRODUCTION

Various health claims have been associated with putative probiotic micro-organisms in both man and animals and three characteristics have been proposed to be desirable for selecting an effective probiotic (Huis in't Veld and Havenaar, 1993). They are:

1. promotion of colonization resistance;
2. influencing of metabolic activities related to host health;
3. stimulation of the host immune response.

Based largely on these criteria, there is now a wide variety of commercial products containing prospective probiotics (mainly lactic acid bacteria) that claim health-promoting effects. Categories (2) and (3) include such purported health benefits as reductions in large bowel carcinogens and mutagens (Gilliland and Speck, 1977), antitumour properties (Reddy *et al.*, 1983; McGroaty, Hawthorn and Reid, 1988), cholesterol-lowering effects (Gilliland, Nelson and Maxwell, 1985), increased lactose digestion (Gilliland and Kim, 1983; Lin *et al.*, 1989), relief from constipation (Graf, 1983), stimulation of immunocompetent cells (Halpern *et al.*, 1991), enhancement of phagocytosis (Perdigon and Alvarez, 1992) and gastrointestinal motility (Friend *et al.*, 1982). These properties have been discussed previously (Fuller, 1992, 1994; Wood, 1992; Tannock 1995) and elsewhere in this book. It is the purpose of this chapter to review the scientific evidence relating to probiotics and improved colonization resistance in humans. This excludes antibiotic-

Probiotics 2: Applications and practical aspects.
Edited by R. Fuller.
Published in 1997 by Chapman & Hall. ISBN 0 412 73610 1

associated diarrhoeal effects (Chapter 3), and focuses on gastroenteritis induced by bacteria or viruses.

Bacteria that produce lactic acid as a major end product of metabolism are the most common commercially available probiotic agents. They mainly include species belonging to the genera *Lactobacillus*, *Pediococcus*, *Leuconostoc*, *Enterococcus* and *Bifidobacterium*. The probiotic concept dictates that these organisms manifest properties that are advantageous towards human health (see below). However, other requirements for an ideal probiotic would include its ability to maintain viability during processing and storage, demonstrable resistance towards the adverse effects of gastric acid and bile, as well as adherence to human intestinal epithelial cells. It is also vitally important that they are completely safe for human consumption (Lee and Salminen, 1995). However, this last criterion has recently been a focus of some debate because some lactic acid bacteria are known to be associated with clinical conditions (Aguirre and Collins, 1993) although these incidents were not related to probiotic administration. While these cases were primarily associated with immunologically compromised patients, the findings nevertheless indicate that considerable care is required in selection of both suitable probiotics and the target population group. Of some relevance to this contention is the recent report by Moore and Moore (1995), where an epidemiological association was made with high faecal counts of bifidobacteria and increased colon cancer risk, although this does not correlate with other apparent antitumour properties of these bacteria (Kohwi *et al.*, 1978; Kohwi, Hashimoto and Tamura, 1982; Sekine *et al.*, 1985). Care should also be taken to avoid using probiotic products that contain micro-organisms that are promiscuous with respect to transfer of genetic information, such as plasmid-borne antibiotic resistance.

Over many years, a consistently important area of probiotic research has been those studies focusing on the suppression of pathogenic activities of micro-organisms in the digestive tract. Some of the scientific evidence obtained in these investigations is reviewed in this chapter, and this is preceded by a discussion on normal homeostasis, the possible mechanisms involved in pathogenesis and the common infections of the gastrointestinal tract.

2.2 HUMAN COLONIC MICROBIOTA AND HOMEOSTASIS

The normal colonic microflora plays an important role in colonization resistance (Hentges, 1983), a term defined as the mechanism whereby the intestinal microbiota protects itself against incursion by new and occasionally pathogenic micro-organisms (Gorbach *et al.*, 1988). Colonization resistance, otherwise known as the barrier effect, bacterial antagonism or bacterial interference, affects homeostasis in the large bowel, and may also be viewed as having a protective role against proliferation of potentially harmful elements present in the autochthonous microbiota.

Although the human large intestinal microbiota has not been fully characterized, we do know that under normal circumstances several hundred different species of bacteria exist in the colon (Moore and Holdeman, 1974; Finegold, Sutter and Mathisen, 1983; Gibson and MacFarlane, 1995). Strictly anaerobic bacteria far outnumber other types of micro-organism, with Gram-negative rods belonging to the *Bacteriodes fragilis* group predominating. Numerically, other Gram-negative organisms such as fusobacteria and enterobacteria constitute a relatively minor proportion of the total faecal flora. Gram-positive rods including bifidobacteria, eubacteria and to a lesser extent, clostridia and lactobacilli, as well as anaerobic Gram-positive cocci including peptococci, peptostreptococci and anaerobic streptococci also inhabit the large bowel. Common probiotic organisms such as the lactic acid bacteria normally constitute a relatively minor component of the gut microbiota (Figure 2.1) although bifidobacteria may comprise as much as 25% of

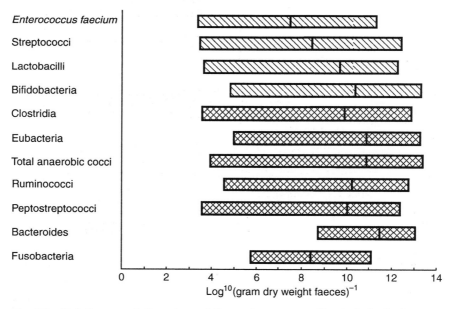

Fig. 2.1 Relative population sizes of the major groups of bacteria in the human large intestine in relation to numbers of bifidobacteria and lactic acid bacteria. Bars show ranges and mean values. Data are from the Wadsworth study with 141 patients (Finegold, Sutter and Mathisen, 1983).

the cultivable gut microflora, with *Bifidobacterium adolescentis* and *B. longum* predominating in adults (Mitsuoka, 1984; Scardovi, 1986), and *Bif. infantis* and *Bif. breve* in infants (Drasar and Roberts, 1990). It is thought that the relatively high proportion of bifidobacteria found in breast-fed infants may be a factor involved in the apparently increased colonization resistance seen in these children (see later). Conversely, the elderly have a relatively low bifidobacterial count and a reduced resistance to infection (Mitsuoka, 1984).

In very general terms, we can categorize the various components of the human gut microbiota into potentially pathogenic or health-promoting groups (Figure 2.2). From the viewpoint of harmful properties, intestinal bacteria may be involved in the onset of localized or systemic infections, intestinal putrefaction, toxin formation and production of mutagenic and carcinogenic substances. Alternatively, some intestinal organisms, for example bifidobacteria and lactobacilli, may

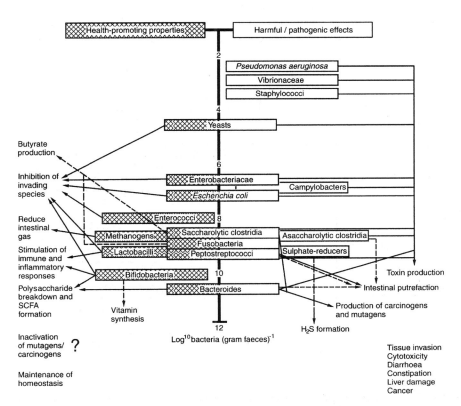

Fig. 2.2 Beneficial, putatively harmful and overtly pathogenic attributes of bacteria growing in the human large intestine. SCFA, short chain fatty acids. Adapted from Gibson and Roberfroid (1995).

confer general health-promoting benefits such as vitamin production, stimulation of the immune system through non-pathogenic means, triglyceride lowering and inhibition of the growth and establishment of harmful microbial species.

Many different factors affect bacterial colonization in the human large intestine, including age, drug therapy, diet, host physiology, peristalsis, local immunity and microbe–microbe interactions (Freter, 1992; Salminen, Isolauri and Onnela, 1995). An important role of the normal flora is related to improved colonization resistance and reduction in the metabolic activities of harmful organisms. A number of mechanisms exist whereby components of the normal gut microbiota can improve colonization resistance. For instance, facultatively anaerobic bacteria maintain a low redox potential (Eh) in the colon by rapidly utilizing traces of oxygen that diffuse into the intestinal lumen. In the newborn, aerobic organisms (coliforms, staphylococci, streptococci) appear within a few days of birth, after exposure to the environment (Drasar and Roberts, 1990). Subsequent growth of these species reduces the Eh, allowing further colonization by anaerobic micro-organisms such as bifidobacteria, bacteroides and eubacteria (Drasar and Roberts, 1990).

Organic acids are the major end products of fermentation in the colon and these metabolites are inhibitory to some invasive bacteria, including species that are potentially pathogenic (Wolin, 1974). It is thought that fermentation lowers intestinal pH to levels where invading species are unable to compete effectively. Acid production, maintenance of a low Eh, as well as the ability to compete for available nutrients and adhesion sites on food particles or at the colonic mucosa, are important factors that determine the composition of the gut flora, with species that are unable to compete being rapidly eliminated from the system. Bacterial interactions that determine whether particular micro-organisms are indigenous to the colon or transient in lumenal contents are unclear. However, some bacteria produce substances that help to protect their particular ecological and metabolic niche.

These products can be distinguished on the basis of whether their activity is primarily against taxonomically related genera, or those whose action is principally against members of the same species. The best known example is the secretion of bacteriocins by *Escherichia coli*. These low molecular weight colicins are effective against other strains of *E. coli*, as well as a small number of other escherichiae and enterobacter (Hill, 1986).

Many investigators have also reported on the abilities of lactic acid bacteria to produce antibacterial substances that are active against pathogenic and food-spoilage organisms (Mehta, Patel and Dave, 1983). Hitherto, two groups of bacterial metabolites have been described as exerting antagonistic effects on other micro-organisms (Geis, 1989):

- bacterial fermentation products, including primary metabolites, such as lactic acid, carbon dioxide, diacetyl, acetaldehyde and hydrogen peroxide (De Vuyst and Vandamme, 1994);
- bacteriocins, which are proteinaceous compounds that manifest anti-microbial activitites against other closely related bacteria.

At least three different groups of bacteriocin are currently recognized (Dodd and Gasson, 1994): the small heat-stable peptides, large heat-labile peptides and modified peptides such as lantibiotics. Lactococci, lactobacilli, pediococci, leuconostocs, carnobacteria, streptococci and enterococci are all known to produce bacteriocins. Reviews by Dodd and Gasson (1994) and De Vuyst and Vandamme (1994) discuss the properties, producer strains, genetic information and target organisms of bacteriocins associated with these bacteria. In addition, bifidobacteria, although strictly speaking not part of the lactic acid group, also secrete antimicrobial agents (Meghrous *et al.*, 1990; Gibson and Wang, 1994; O'Riordan, Condon and Fitzgerald, 1995). From the viewpoint of the human large intestine, and their widespread probiotic usage, antimicrobial substances formed by lactobacilli and bifidobacteria will be considered further.

Lactocin 27, produced by *Lactobacillus helveticus* LP27, is a heat-stable lipopolysaccharide protein complex that is bacteriostatic towards other lactobacilli (Upreti and Hindsdill, 1975). Another strain of *L. helveticus* (487) produces helveticin J, a large heat-labile bacteriocin (Klaenhammer *et al.*, 1992), while *L. helveticus* CNRZ450 secretes a small complex bacteriostatic protein of about 30–50 kDa, which is active against a narrow range of closely related organisms (Thompson *et al.*, 1996). Two species of *L. acidophilus*, N2 and 11088, produce small heat-stable antimicrobial substances known as lactacins B and F respectively (Barefoot, Nettles and Chen, 1994; Klaenhammer, Ahn and Muriana, 1994), while curvacin C has been isolated from *L. curvatus* LTH 1174. This bacteriocin causes lysis of other lactobacilli, as well as food contaminants such as carnobacteria and listeria (Tichaczek *et al.*, 1992). Three species of *L. sake* (L45, LB706, LTH673) form the bacteriocins lactocin S, sakacin A and sakacin P respectively. The former is a lantibiotic (Nes *et al.*, 1994), whilst the sakacins are small heat-stable bacteriocins (Tichaczek *et al.*, 1992; Schillinger, 1994). Caseicin 80 from *L. casei* B80 is a heat-labile bacteriocin with a relatively narrow activity spectrum, in that it affects other strains of *L. casei* (Rammelsberg and Radler, 1990). Antimicrobial substances produced by *L. plantarum*, such as plantaricin SIK-83, affect a wide range of lactic acid bacteria including enterococci, streptococci and pediococci, as well as *Staphylococcus aureus* (Andersson, 1986). Itoh *et al.* (1995) showed that gassericin A, which is produced by *L. gasseri*, is bacteriocidal but not bacteriolytic to *Listeria monocytogenes* and other enteric pathogens. This bacteriocin is a small peptide with a molecular mass of about 3.8 kDa (Kawai *et al.*, 1994).

Lactobacillus reuteri is an important member of the lactobacillus population in the human gastrointestinal tract. This organism produces reuterin, a non-proteinaceous antimicrobial agent that exhibits a broad inhibitory activity affecting Gram-negative (salmonellae, shigellae) and Gram-positive (clostridia, listeria) bacteria, fungi and protozoa (Axelsson *et al.*, 1989).

Bifidobacteria also produce antagonistic substances that inhibit growth of other bacteria (Tamura, 1983; Araya-Kojima *et al.*, 1995). The mechanism whereby this occurs has been largely attributed to production of fermentation acids (acetate, lactate) during growth. However, Anand, Srinivasan and Rao (1985) and Mantere-Alhonen, Noro and Sippola (1989) have shown that bifidobacteria (*Bifidobacterium bifidum*, *B. longum*) may also be inhibitory in a manner that is unrelated to culture pH. This has been confirmed by Gibson and Wang (1994) in a series of studies summarized below.

(a) Co-culture experiments

Fermentation vessels inoculated with pure cultures of *Bif. infantis*, and the pathogenic organisms *E. coli* and *Clostridium perfringens* were operated statically. Initially, all three species were added to the same fermenter which was controlled at neutral pH. Viable counts of *C. perfringens* and *E. coli* declined markedly during the incubation, while populations of *Bif. infantis* remained high throughout the experiment.

(b) Diffusion chemostat

This system facilitated growth of co-cultures in two different fermentation vessels separated by a semi-permeable membrane filter, which allowed diffusion of various growth factors and bacterial metabolites, but no bacterial cells. In these studies, *C. perfringens* and *E. coli* numbers decreased when the pH was lowered from 7.0 to 5.3; however, the inhibitory effect was enhanced during co-culture with *Bif. infantis* NCFB 2205.

(c) Inhibitory effects of bifidobacteria on other micro-organisms

Batch culture experiments showed that *Bif. infantis* cell-free extracts were inhibitory towards *Bact. fragilis*, *E. coli* and *C. perfringens*. This was demonstrated in comparisons of bacterial growth curves in the presence of sterilized culture media and *Bif. infantis* cell-free culture supernatants. In this case, the inhibitory effect could neither be attributed to competition for growth substrates (the supernatant was sterile) nor low pH (the extract pH was 6.5).

(d) Chemostat fermentations

Continuous culture experiments showed that in fermenters maintained at pH 7, addition of an overnight culture of *Bif. infantis* was inhibitory to actively growing populations of both *E. coli* and *C. perfringens* each showing a $0.5 \log_{10}$ decrease compared to growth in the absence of the bifidobacterium.

After incubation of plate cultures of both *C. perfringens* and *E. coli* clear zones, where bacterial growth had been inhibited, were evident around filter paper discs that had been previously soaked in a culture of *B. infantis*. Sterile growth media used as a replacement for the bifidobacterium did not produce this effect. Repeat experiments in which methanol–acetone partially purified extracts (M-A) replaced the *B. infantis* culture caused a more marked inhibitory effect. HPLC analysis showed that the extract did not contain acetic or lactic acids.

(e) Effect of bifidobacteria on pathogenic micro-organisms

M-A extracts from eight different species of bifidobacteria (*Bif. infantis* NCFB 2205, *Bif. longum* NCFB 2259, *Bif. pseudolongum* NCFB 2244, *Bif. catenulatum* NCFB 2246, *Bif. bifidum* NCFB 2203, *Bif. breve* NCFB 2257, *Bif. adolescentis* NCFB 2230, *Bif. angulatum* NCFB 2238) directly inhibited growth of a range of pathogenic bacteria, including species belonging to the genera *Salmonella, Shigella, Listeria, Escherichia, Vibrio, Campylobacter, Clostridium* and *Bacteroides*. The degree of antibacterial activity was variable, with the most potent effects generally being exerted by *Bif. infantis* and *Bif. longum*.

The antimicrobial activities of bifidobacteria have not been well characterized, although they appear to exhibit a much wider activity spectrum than is usually associated with conventional bacteriocins. Studies by O'Riordan, Condon and, Fitzgerald (1995) indicate that *Bif. infantis* NCFB 2255 and *Bif. breve* NCFB 2258 may produce two different types of antimicrobial agent. One is thought to be proteinaceous in nature and largely affects Gram-positive bacteria, whilst the other was not affected by proteolytic enzymes, and inhibited Gram-negative organisms.

Although laboratory studies demonstrate the antimicrobial activities of lactobacilli and bifidobacteria, it is unclear whether production of antagonistic substances is of any real ecological significance in the human colonic microbiota. However, these activities may help to explain the apparent effects that some probiotic preparations have on symptoms of gastroenteritis.

2.3   INFECTIONS OF THE INTESTINAL TRACT

In man, acute enteritis or colitis can result from infection by a variety of micro-organisms, including fungi, protozoa, viruses and bacteria. Although viral forms of enteric disease are most frequently diagnosed, large intestinal infection by bacteria is also a serious clinical problem, and many bacterial pathogens are of worldwide significance. Figure 2.3 shows the principal human intestinal infections mediated by bacteria and viruses, of which a number have no identifiable aetiologic agent.

In general terms, bacteria growing in the large intestine can be differentiated into three broad categories: autochthonous micro-organisms, which are true inhabitants of the colon that have undergone evolution with the host; indigenous bacteria that at some stage have been able to colonize the gut; and contaminants, which are micro-organisms that reside in the intestine for short periods, and do not permanently establish in the ecosystem. The majority of infectious agents belong to the third group and, occasionally, homeostasis is perturbed by these invasive micro-organisms resulting in severe diarrhoeal disease.

Human intestinal pathogens manifest a wide variety of virulence factors that enable them to overcome host defences (Cohen and Giannella, 1991). Invasive micro-organisms multiply within enterocytes or colonocytes and ultimately cause cell death. Examples include enteroinvasive *E. coli*, salmonellae, shigellae, yersiniae, campylobacters, vibrios and aeromonads. Other pathogens such as enteropathogenic (EPEC) and enterohaemorrhagic (EHEC) strains of *E. coli* are cytotoxic.

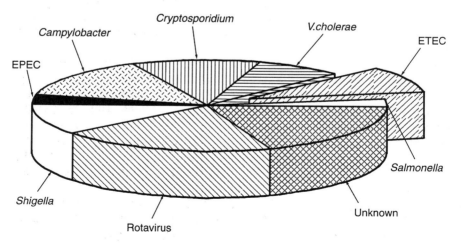

Fig. 2.3 Predominant infections of the human intestinal tract. Data are from World Health Organization figures in 1990. EPEC, enteropathogenic *E. coli*; ETEC, enterotoxigenic *E. coli*.

They secrete a variety of agents that directly cause cell injury. Many pathogens are of course invasive and toxigenic. Toxigenic micro-organisms, including *Vibrio cholerae*, some shigellas and enterotoxigenic *E. coli* (ETEC), secrete enterotoxins that affect intestinal salt and water balance. Other bacteria that cause acute inflammation in the intestine are adhesive micro-organisms, such as ETEC and enteroaggregative *E. coli* (EAggEC) that bind tightly to the gut mucosa.

During evolution, the human host has evolved a variety of effective defence mechanisms against invading micro-organisms. The major non-immunological host-defence processes involve gastric acid secretion, intestinal motility, lysozyme, pancreatic secretions and bile. Mucins are constantly secreted by the gastrointestinal tract, and these glycoproteins form an efficient physical barrier against invasion and colonization (Quigley and Kelly, 1995).

The final major defence against infection of the large bowel is afforded by the autochthonous flora. This resistance to newly intro-duced bacteria is well known, and is responsible for the long-term stability of the gut microbiota. The barrier effect is classically seen in animal studies, where germ-free animals are more susceptible to orally administered shigellae and salmonellae compared with their conven-tional counterparts. However, it has also been convincingly demon-strated in humans, where patients with *Clostridium difficile*-associated colitis have been successfully treated by rectal instillation of faecal suspensions from healthy donors (Bowden, Mansberger and Lykins, 1981; Schwan *et al.*, 1984; Tvede and Rask-Madsen, 1989). While it is generally thought that the barrier effect of the normal gut microbiota is a property largely conferred by its obligately anaerobic bacterial compo-nents, Gorbach *et al.* (1988) showed that this was not necessarily the case, when colonization resistance was eliminated by antibiotic therapy specific for aerobes. The early studies of Sears *et al.* (1956) lend some support to this argument; they used a non-virulent *E. coli* to protect a dog from invasion by a virulent strain of the bacterium.

Often, however, the host defence becomes compromised or overcome by pathogenic micro-organisms, resulting in infection. This may arise because of toxins elaborated by the pathogen, environmental factors, age, stress or associated underlying illness in the host.

2.4 ATTACHMENT

The adherence of micro-organisms to epithelial surfaces is one of the most important initiating events in pathogenesis in many, though by no means all, infections of the gastrointestinal tract. Adherence of pathogens to mucosal surfaces in the gastrointestinal tract assists in preventing their removal by host secretions and lumenal flow. Rapid growth is an important factor in the ability of micro-organisms to colonize epithelial surfaces, and bacterial attachment to epithelia is mediated by a diverse range of surface structures or molecules collectively known as adhesins. They recognize specific epithelial surfaces (MacFarlane and Gibson, 1995), intestinal mucus (Cohen, Wadoloski and Laux, 1986), as well as intercellular structures (Savage and Fletcher, 1985). Many intestinal pathogens produce multiple adhesins such as the fimbrial adhesins (CFA I, CFA II, K88 and K99) in *E. coli* (Smith, 1995).

As discussed earlier, the abilities of non-pathogenic bacteria to colonize epithelial surfaces, thereby excluding pathogenic species, is considered to be an important probiotic trait. Bifidobacteria and lactobacilli have been shown to be adherent to human and animal epithelial cells. A review of the literature suggests that these organisms form a variety of adhesins. For example, Op den Camp, Oosterhoh and Veerkamp (1985) suggested that hydrophobic interactions mediated by lipoteichoic acids might be important in the adherence of bifidobacteria to intestinal epithelial cells, whereas Sato, Mochizuki and Homma (1982) concluded that *Bif. infantis* adhered to porcine ileal epithelial cells, and that binding was enhanced by production of extracellular polysachharide by the organism.

In vitro studies with human cell lines have shown that bifidobacterial isolates of human origin adhered to Caco-2 cells, and inhibited interactions between these cells and enteropathogens such as *Yersinia pseudotuberculosis, Salmonella typhimurium*, EPEC and ETEC (Bernet *et al.*, 1993). Bifidobacteria adhere at the apical brush border without causing cell damage, with cell binding facilitated by the formation of species-specific extracellular and cell-associated protein-like adhesins by the bacteria. *Bifidobacterium infantis* and some strains of *Bif. breve* and *Bif. longum* attached strongly, although other *Bif. breve* and *Bif. longum* isolates were poorly adherent, demonstrating species and strain variations in expression of this probiotic attribute.

Other *in vitro* studies also showed that bifidobacteria vary greatly in their abilities to adhere to human colonic epithelial cells, at species and strain level (Figure 2.4), and that this observation correlated with their capacities for large intestinal colonization *in vivo* (Crociani *et al.*, 1995). These authors related differences in adhesive traits to variations in bifidobacterial cellular proteins, polysaccharides and lipoteichoic acids, as well as surface ionic charge.

Although not all lactobacilli are able to attach to human intestinal epithelial cells (Kleeman and Klaenhammer, 1982), species that do colonize the gut in this way characteristically exhibit high surface hydrophobicities (Wadstrom *et al.*, 1987). Adherence of lactobacilli to epithelial cells appears to be facilitated by secretion of a proteinaceous substance (Chauviere *et al.*, 1992b). This is supported in part by the

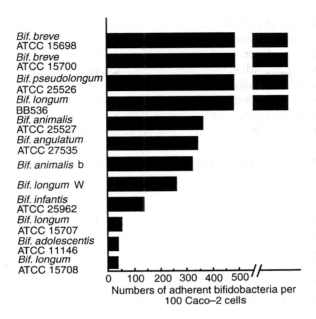

Fig. 2.4 Comparison of attachment frequencies of different species and strains of bifidobacteria to human enterocytes *in vitro*. Data are modified from Crociani *et al*. (1995).

studies of Reid *et al*. (1993), who observed that three different strains of *L. casei* adhered to human uroepithelial and colonic epithelial cells using an extracellular protein-like adhesin. However, these authors also identified a cell-wall-associated adhesin that was insensitive to protease treatment. Conway, Gorbach and Goldin (1987) reported that adhesion of lactobacilli was non-specific and that the organisms attached to pig and human epithelial cells.

Colonization of the human intestinal mucosa by lactobacilli has been demonstrated by Johansson *et al*. (1993). They administered a dose of 19 test strains, belonging to several species, in oatmeal soup to a group of 13 healthy volunteers. Jejunal biopsies showed that lactobacilli increased, while numbers of streptococci and clostridia declined on the mucosal surface during feeding of the probiotic. Significant reductions were also seen in anaerobe populations on the rectal mucosa and, in some volunteers, enterobacterial numbers decreased 1000-fold. The authors concluded that this was due to lactobacillus antagonism, as adminstered *L. plantarum* (two strains), *L. casei* subsp. *rhamnosus*, *L. reuteri* and *L. agilis* were isolated from the rectal mucosa. These organisms were persistent, since they were present for up to 11 days after probiotic feeding had ceased. Administration of lactobacilli has been reported by other workers to reduce faecal excretion of *E. coli* in humans (Lidbeck, Gustafsson and Nord, 1987).

Lactobacilli have also been shown to competitively exclude diarrhoeagenic *E. coli* from human epithelial cells in *in vitro* tests (Chauviere *et al.*, 1992a) and to inhibit adhesion of pathogenic *E. coli* to mucin (Blomberg, Hentiksson and Conway, 1992). Whole cells may not be necessary to competitively exclude pathogenic micro-organisms; Chan *et al.* (1985) demonstrated that attachment of Gram-negative pathogens to uroepithelial cells was inhibited if the eukaryotic cells had been preincubated with cell fragments of a lactobacillus isolated from the urinary tract. More recent work has also shown that low molecular mass (*c.* 1.7 kDa) components of the cell wall, released during lysis of *L. fermentum*, inhibited adhesion of enterotoxigenic *E. coli* K88 to porcine ileal mucus (Ouwenhand and Conway, 1996).

2.5   USE OF PROBIOTICS AGAINST INTESTINAL INFECTIONS

The literature contains a plethora of seemingly contradictory results concerning the efficacy of various probiotic preparations used in preventing infection in the intestinal tract. There are many potential reasons for this; however, natural interindividual variations in the test subjects or patients, together with species and strain differences in the probiotics are of particular importance (Conway and Henriksson, 1994). When probiotic treatments have seemingly been successful, high daily doses of the putative probiotic agent appear to have been required for beneficial effects to be manifested and maintained. Culture viability should therefore be viewed as an important factor in probiotic selection by the consumer. However, a recent study (Hamilton-Miller, Shah and Smith, 1996) investigated the species composition and viability of 13 commercially available (over the counter) probiotic preparations in the UK; it was shown that only two of the products contained the micro-organisms claimed on the labels, with respect to bacterial type and viable cell numbers. The other preparations either contained organisms that were not listed on the product labels, did not contain the bacteria stated on the packaging, or the bacteria were present, but in considerably lower numbers than claimed.

Lactic acid and bifidobacterial probiotic effects include biochemical, physiological and antimicrobial activities on either autochthonous or allochthonous elements of the colonic microbiota. Rolfe, Helebian and Finegold (1981) found that in *in vitro* studies where 401 faecal bacteria from 23 different genera were investiaged, only *Bif. adolescentis, Bif. infantis, Bif. longum*, two unspeciated bacteroides, two unspeciated lactobacilli and a number of enterococci exhibited direct inhibitory activity against *Clostridium difficile*, the principal aetiologic agent of pseudomembranous colitis. It was also found that bacterial interference was strongly influenced by the type of culture medium used. Competition studies with faecal bacteria demonstrated that availability of glucose, N-acetylglucosamine and sialic acids was a significant factor affecting the ability of *C. difficile* to colonize the gut (Wilson and Perini, 1988).

Various groups have used lactobacilli in the treatment of pseudomembranous colitis, with mixed results. Aronsson, Barany and Nord (1987) found that administration of freeze-dried powders of *L. acidophilus* NCDO 1748 had no effect on patients with *C. difficile*-associated colitis; however, subsequent studies with *Lactobacillus* GG successfully eradicated this pathogen in a group of five patients with relapsing *C. difficile* colitis (Gorbach, Chang and Goldin, 1987). The lactobacillus had previously been shown to colonize the gut and secrete an antimicrobial product active against *C. difficile* and a range of other micro-organisms (Silva *et al.*, 1987). Approximately 10^{10} viable bacteria were fed daily to the patients in 5 ml skimmed milk. Diarrhoeal symptoms were immediately relieved in four individuals, with concomitant reductions in *C. difficile* toxin titres in stools. The other patient also improved following further antibiotic and probiotic treatment.

The non-pathogenic yeast *Saccharomyces boulardii* has been used in prevention and treatment of antibiotic-associated diarrhoea caused by *C. difficile* (Surawicz *et al.*, 1989 a, b). In a group of 180 patients taking part in a double-blind controlled study, 9.5% of those receiving the probiotic manifested diarrhoeal symptoms compared to 22% in controls receiving a placebo. The authors concluded that prophylactic use of the probiotic reduced the incidence of diarrhoea associated with *C. difficile* infection, and that no adverse effects in the patients could be ascribed to *Sac. boulardii*. Although *Sac. boulardii* was shown not to prevent acquisition of *C. difficile*, the mechanism whereby the yeast prevented diarrhoea was not established.

Prophylactic use of lactic acid bacteria preparations has been investigated in relation to travellers' diarrhoea. Clements *et al.* (1981) carried out trials with lactobacilli aimed at preventing travellers' diarrhoea caused by ETEC. The prospective probiotic consisted of approximately 10^8 bacteria in 0.24 L skim milk, and was fed to volunteers 36 h before, and up to 96 h after administration of the enterobacterial pathogen. The authors found no real differences in symptoms in the control and test groups. In a later study, Black *et al.* (1989), using a mixture of *Bif. bifidum, Streptococcus salivarius* ssp. *thermophilus, Lactobacillus delbrueckii* ssp. *bulgaricus* and *L. acidophilus*, reduced the incidence of travellers' diarrhoea by nearly half in visitors to Egypt, while *Lactobacillus* GG given as a powder (c. 10^9 viable bacteria per day) reduced the incidence of diarrhoea in one of two groups of 756 people tested (Oksanen *et al.*, 1990).

In studies on the use of probiotics to treat specific illnesses other than C. difficile-associated colitis, Pearce and Hamilton (1974) gave 94 children a mixture of S. salivarius ssp. thermophilus, L. delbrueckii ssp. bulgaricus and L. acidophilus in powder form, providing 10^8–10^9 bacteria per day. No alleviating effects were observed with respect to the onset of acute diarrhoea. However, many other groups have reported on the successful use of bifidobacteria and lactic acid bacteria to treat diarrhoeal illness. For example, Enterococcus faecium has been useful in treating enteritis in adults (Camarri et al., 1981) and acute diarrhoea in infants (D'Appuzo and Salzberg, 1982). Hotta et al. (1987) used a commercially available bifidus yoghurt and oral preparations of Bif. breve to treat paediatric intractable diarrhoea in 15 patients; they reported a significant improvement in all subjects, 3–7 days after probiotic administration. Tojo et al. (1987) observed that oral administration of B. breve eradicated Campylobacter jejuni from stools of children with enteritis, although this occurred less rapidly than in patients treated with erythromycin.

In their study, Benno and Mitsuoka (1992) gave 3 g of a Bif. longum preparation (10^9 viable cell g^{-1}) to five healthy volunteers for 5 weeks. This had no significant effect on numbers of the major groups of intestinal bacteria excreted in faeces, although counts of lecithinase-negative clostridia were reduced. However, β-glucuronidase activities and faecal ammonia were lowered during the study, together with faecal pH in the last week of feeding. Feeding people bifidobacteria in dairy products does not always result in demonstrable changes in the colonic microbiota composition or activities. Bartram et al. (1994) compared the effects of ordinary yoghurt and a synbiotic yoghurt enriched with B. longum ($> 10^9$ viable cell ml^{-1}) and lactulose (5 g 1^{-1}) on a variety of faecal parameters, in a double-blind crossover study with 12 healthy volunteers. With the exception of an increase in faecal excretion of bifidobacteria, no significant changes were observed in either gut bacteriology, intestinal transit time, stool weight, faecal pH, short-chain fatty acids, bile acids or neutral sterols.

2.6 PROBIOTICS AND VIRAL INFECTIONS OF THE GASTROINTESTINAL TRACT

The intuitive notion that ingestion of beneficial bacteria helps to improve homeostasis and protects against invasion of the gastrointestinal tract by harmful micro-organisms is probably an oversimplification of the inhibitory mechanisms exerted by probiotics. An example of this is the effect that probiotics may have on the prevention or treatment of gastrointestinal bacterial infections (see earlier). With respect to viruses, similar inhibitory mechanisms to those exerted on bacteria may exist, such as direct inhibitory effects akin to those seen with bacteriocins, although few inhibitory substances produced by bacteria have been described that target viruses. Changes in the gastrointestinal milieu such as modification of Eh, competition for nutrients or exclusion of attachment sites are possible, but they are unlikely to inhibit multiplication of viral pathogens.

However, there is a strong possibility that viral infections in the gastrointestinal tract may be prevented through immunomodulatory mechanisms exerted by probiotics. These mechanisms are discussed elsewhere in this book, but we will describe examples that are specifically relevant to viruses. A major immunological barrier against viruses is secretory IgA. Studies in mice have shown that *Lactobacillus casei* and yoghurt administration to BALB/c mice resulted in a dose-dependent increase in IgA-secreting B cells in GALT (gut associated lymphoid tissue) of the small and large intestine (Perdigon *et al.*, 1994, 1995). Using a murine Peyer's patch cell culture method, Yasui *et al.* (1992) demonstrated that three different strains of bifidobacteria, when administered orally with a cholera toxin, induced a significant increase in antitoxin IgA antibody production and proliferation of Peyer's patch cells.

The mechanisms for this immunomodulation are unclear. Certain *in vitro* studies indicate the possibility of cytokine production which may mediate immunostimulation. *Bifidobacterium breve* added to mouse Peyer's patch cells cultured with sheep red blood cells induced proliferation of B cells and also elicited an enhanced anti-lipopolysaccharide antibody production (Yasui and Ohwaki, 1991). Solis and Lemmonnier (1993) reported that several agents, including bifidobacteria, *L. acidophilus*, *L. casei*, *L. helveticus* and yoghurt, when incubated with blood mononuclear cells from healthy adults, produced varying increases in interleukin 1-B, tumour necrosis factor alpha and gamma interferon. More recently, Kitazawa *et al.* (1994) carried out *in vitro* stimulation by *L. gasseri* of spleen macrophages and adherent cells from Peyer's patches of BALB/c mice and reported increased production of mRNA-mediated alpha interferon.

Animal studies have shown that several probiotics may have a beneficial effect against rotaviral diarrhoea. Duffy *et al.* (1994a) showed a delayed clinical onset, lower prevalence and decreased rotaviral shedding in murine viral diarrhoea in mice fed with *Bif. bifidum*. Viral shedding was also reduced (Duffy *et al.*, 1994b). An interesting and provocative study by Yasui, Kiyoshima and Ushima (1995) demonstrated passive protection against rotaviral induced diarrhoea in mouse pups born to and nursed by dams fed *B. breve*. These investigators demonstrated, in this well-controlled study, that the protection was mediated through an increase in specific antirotaviral IgA in maternal milk. In mammals, including humans, the entero-mammary pathway for secretory IgA production is thought to be an important protective mechanism against infections in breast-fed children. When the gut and respiratory systems are exposed to specific antigens, such as bacteria or viruses, this leads to stimulation of precommitted B lymphocytes that undergo an isotype switch from IgM to IgA. These cells leave either the enteric or respiratory system via efferent lymphatics, enter the blood-stream and reach the mammary gland. Once in the gland, the B lymphocytes transform into plasma cells which produce specific IgA that is subsequently secreted into the milk, thus providing specific passive protection to the gut of the infant, lightly exposed to the same agent as the mother.

Human studies are beginning to validate similar mechanisms and clinical effects. Link-Amster *et al.* (1994) demonstrated a four-fold rise in specific serum IgA against an attenuated *Salmonella typhi* given to volunteers who consumed *L. acidophilus* and bifidobacteria over a 3-week period. An increase in alpha interferon and B-cell frequency after ingestion of *B. bifidum* and *L. acidophilus* by healthy adults has been demonstrated by De Simone *et al.* (1992).

Clinical studies in children have also been encouraging. In one study, Isolauri *et al.* (1991) randomized 71 well-nourished children with acute diarrhoea to receive *L. casei* strain GG and strain GG fermented milk product or pasteurized yoghurt as a placebo. Both study groups showed a significant shortening of the duration of the diarrhoea (1.4 vs. 2.4 days). The clinical significance of this difference is marginal. However, the same group of investigators (Kaila *et al.*, 1992) demon-strated that at convalescence, 90% of a group treated with lactobacilli, against 45% of a placebo group, developed specific IgA antibody-secreting cell response to rotavirus.

Since priming of GALT may be an important part of this protective mechanism, some work has focused on the preventative effects of regular probiotic consumption against rotaviral disease. In a recent double-blind placebo-controlled trial, randomized infants less than 24 months of age, who were admitted to a chronic medical care facility, received standard infant formula or formula supplemented with *Bif. bifidum* and *S. salivarius* ssp. *thermophilus* (Saavedra *et al.*, 1994). Thirty one per cent of patients in the control group, but only 7% of those in the supplemented group developed acute diarrhoea. Moreover, only 10% of the supplemented group against 39% of the controls shed rotavirus in stools at some time during the 17-month study. Further work with orally administered *L. casei* given in conjunction with rhesus-human reassortant live oral rotavirus vaccine to infants aged 2–5 months resulted in an elevated response with regard to rotavirus-specific IgM-secreting cells, and improved anti-rotavirus IgA serocon-version (Isolauri *et al.*, 1995).

This work indicates the potential for use of probiotic agents in the prevention and treatment of acute viral diarrhoea, particularly that caused by rotavirus – the form of diarrhoea that continues to be the most significant worldwide. Nevertheless, caution should be exerted in generalizing these results, since there are still many inherent difficulties involved in interpreting the information obtained: the variability between studies and the agents used, the dose and dose regimes and, importantly, the specific clinical outcomes (Saavedra, 1995).

2.7 MODULATION OF THE HOST RESPONSE TO INFECTION

Together with competition for binding sites on epithelial cells and secretion of antagonistic substances, lactobacilli and bifidobacteria have been shown to specifically influence the host response to infection. *Bifidobacterium bifidum* and *L. acidophilus*, given in two capsules, four times daily for 28 days, reduced colonic inflammatory infiltration in 15 elderly subjects (De Simone *et al.*, 1992). An increase in B-cell numbers was also observed in peripheral blood, leading the authors to conclude that regular oral bacteriotherapy using these organisms resulted in modulation of both immune and inflammatory responses in their subjects. The physiological mechanisms for these phenomena are unclear; however, the authors considered that the probiotics exerted a 'barrier effect' against colonic pathogens or their secretory products that were responsible for localized inflammatory responses in the gut. Proliferation of peripheral B lymphocytes was attributed to antigen-sensitized cells entering the general circulation from efferent lymphatic vessels.

Yasui *et al.* (1992) screened 120 bifidobacterial strains isolated from human faeces for their abilities to stimulate IgA production. Two *Bif. breve* strains and one *B. longum* induced high sIgA activities. Oral adminstration of one of the *B. breve* isolates to mice was found to augment the immune response to cholera toxin in GALT.

Kaila *et al.* (1992) reported that oral introduction of lactobacilli increased intestinal IgA production during diarrhoea in humans. However, not all lactobacilli are immunostimulatory, and for those that are, the size of the administered dose seems to be critical. In studies with mice, Perdigon *et al.* (1993) found that oral administration of *L. casei*, but not *L. acidophilus* induced synthesis of IgA, and increased the magnitude of the inflammatory response. *Lactobacillus casei* exerted an adjuvant effect in IgA synthesis in the gut, and long-term feeding of the bacterium increased proliferation of plasma cells, lymphocytes and macrophages.

2.8 EFFECTS OF PROBIOTICS IN INFANTS

It is claimed that differences in the faecal microfloras of breast-fed and bottle-fed infants are responsible for the seemingly lower risk of infection enjoyed by the former group. In this context, bifidobacteria are one of the dominant organisms in the gut of breast-fed infants, outnumbering other bacteria by at least 100-fold (Braun, 1981; Benno, Sawado and Mitsuoka, 1984; Modler, McKellar and Yaguchi, 1990; Yoshita *et al.*, 1991). A number of explanations have been put forward for the dominance of bifidobacteria in breast-fed infants. For example, human breast milk contains growth factors stimulatory for these organisms, such as N-acetylglucosamine, glucose, galactose and fucose (Gyorgy, 1953; Gauhe *et al.*, 1954). It is also thought that the low protein content with resultant reduced buffering capacity of human milk facilitates increased bifidobacterial growth (Willis *et al.*, 1973; Bullen *et al.*, 1977; Faure *et al.*, 1984), with the presence of lactoferrin also being a contributory factor (Drasar and Roberts, 1990). As a result, attempts have been made to modify infant feed formulae to more closely resemble that of humans, including adjustments to the protein content (Modler, McKellar and Yaguchi, 1990), and the use of probiotics. An early report by Robinson and Thompson (1952) showed that supplementation of artificial milk with lactobacilli improved weight gain in formula-fed infants, but had no effect in those that were breast fed.

Other studies have investigated probiotic supplementation in an attempt to reduce the symptoms of childhood diarrhoea. Both lactobacilli and bifidobacteria have been used, especially strains of *L. acidophilus* (Beck and Necheles, 1961) and *L. casei* (Isolauri *et al.*, 1991). Oral administration of *L. acidophilus* was shown by Zychowicz *et al.* (1974) to reduce the carrier state of salmonella and shigella in children, and similar affects on salmonellae were found by Alm (1983). Tomic-Karovic and Fanjek (1962) showed that acidophilus milk could be used to treat *E. coli*-associated diarrhoea. In contrast, however, Pearce and Hamilton (1974) concluded that oral dosing of lactobacilli had no significant effect on diarrhoeal disease in children. Differences in controls used in these studies, and the general scientific design of the trials, for example whether they were carried out double blind or not, may account for apparent inconsistencies. This, together with variations in selection criteria, probably explains a great deal of the variability seen in data associated with probiotic therapies.

Attempts have been made to use non-pathogenic forms of certain organisms to induce bacterial interference against the virulent forms. Examples include the studies by Shinefield, Ribble and Boris (1971) who worked with *Staphylococcus aureus*, and Rastegar Lari *et al.* (1990) for *E. coli*. In these investigations, the authors concluded that the non-pathogenic variants occupied colonization sites, thereby excluding the harmful form.

For bifidobacteria, use of a dietary supplement containing *Bif. bifidum* reduced infections in a large group of infants (114) by approximately eight-fold (Kaloud and Stogmann, 1969). The placebo was buttermilk.

Other applications of the same organisms in infants have been success-fully used in reducing overgrowth of *Candida albicans* in the gut after antimicrobial therapy (Mayer, 1969) and for treating enterocolitis (Tasvac, 1964).

2.9 PREBIOTICS AND SYNBIOTICS

2.9.1 Health aspects of the normal gut microflora

In humans, colonic fermentation may improve some aspects of host health. As discussed earlier, the normal gut microflora contains bacterial populations that can be perceived as being health promoting, as well as being pathogenic (Figure 2.2). To augment these beneficial components of the microbiota, dietary supplementation has been suggested (Fuller, 1989).

With the consideration that many potentially health-promoting micro-organisms such as bifidobacteria and lactobacilli are already resident in the human colon, Gibson and Roberfroid (1995) have introduced the prebiotic concept. A prebiotic is a non-digestible food ingredient that beneficially affects the host by selectively stimulating the growth and/or activity of one or a limited number of bacteria in the colon, that can improve the host health. For a food ingredient to be classified as a prebiotic, it must:

- neither be hydrolysed nor absorbed in the upper part of the gastrointestinal tract;
- be a selective substrate for one or a limited number of potentially beneficial bacteria commensal to the colon, which are stimulated to grow and/or are metabolically activated;
- consequently, be able to alter the colonic microflora towards a healthier composition, for example by increasing the number of saccharolytic species and reducing putrefactive micro-organisms such as asaccharolytic clostridia.

Any food ingredient that enters the large intestine is a candidate prebiotic. However, to be effective, selective fermentation by the colonic microbiota is required. This may occur with non-digestible complex carbohydrates, some peptides and proteins, as well as certain lipids. Because of their chemical structure, these compounds are not absorbed in the upper part of the gastrointestinal tract. Non-digestible carbohydrates, in particular fructose-containing oligosaccharides, are naturally occurring in a variety of plants such as onion, asparagus, chicory, banana and artichoke, and fulfil the prebiotic criteria. The average daily Western-style diet contains approximately 5 g of fructo-oligosaccharides (Van Loo *et al.*, 1995).

Data from *in vitro* studies with batch culture fermenters demonstrated that fructo-oligosaccharides were specifically fermented by bifidobacteria (Wang and Gibson, 1993). This was subsequently confirmed in a human volunteer trial using both oligofructose and inulin, at a level of 15 g per day (Gibson *et al.*, 1995), where feeding fructo-oligosaccharides to healthy human volunteers caused bifidobacteria to become the numerically predominant bacterial genus in faeces. The volunteers' diet was strictly controlled apart from the use of a placebo carbohydrate. Thus, the bacterial changes that were seen could be ascribed to the fructo-oligosaccharide being metabolized in the large intestine. Bearing in mind the types of foods that have high fructo-oligosaccharide contents, it is probably not realistic to advocate increased intake to the levels required for effective flora manipulation – although a dose response study has not been carried out. It is probably more feasible to add purified forms of the oligosaccharides to commonly ingested foods such as dairy products, biscuits or breakfast cereals.

Other non-digestible oligosaccharides that may have prebiotic potential include raffinose and stachyose, as well as those that contain xylose, galactose and maltose (Rumney and Rowland, 1995). In particular, feeding 5% (w/v) galacto-oligosaccharides to human microflora-associated rats has been shown to significantly increase populations of bifidobacteria and lactobacilli, while decreasing enterobacteria (Rowland and Tanaka, 1993).

Bifidogenic effects of oligosaccharides have also been reported by Japanese workers (Yawaza, Imai and Tamura, 1978; Yawaza and Tamura, 1982; Hidaka *et al.*, 1986; Mitsuoka, Hidaka and Eida, 1987; Hayawaka *et al.*, 1990). However, the specificities of these effects were not investigated with respect to proliferation of other components of the gut microbiota.

Although it remains to be demonstrated that prebiotics have any direct remedial role in human health, the observation that the human gut microflora composition can be manipulated by diet does have potential health implications. Possible applications are similar to those of probiotics and include improved products for the food industry, design of novel products, as well as improvement of existing foods and prophylactic management of gastrointestinal disease that involves modulating the composition or metabolic activities of the gut microflora. The advantage of prebiotics is that the target micro-organism(s) is already commensal to the gut, occupying an effective colonization site and should not cause immunological problems that could possibly be associated with intake of foreign antigens. By careful consideration of the gut microflora composition and activities, it should be possible to develop prebiotics for specific purposes.

Tannock (1995) has summarized the potential use of molecular biological techniques for improved probiotic efficacy. Similarly, the application of molecular-based methodologies such as genetic fingerprinting, nested probing and restriction fragment length polymorphism, which more adequately determine the microflora composition, should also add positively to the potential of prebiotics. Most gut microbiologists agree that our knowledge of the full diversity of the colonic microflora is far from complete. Recent advances in the use of 16S rRNA technology combined with the polymerase chain reaction as used for various ecosystems may now be applied to the human colon and, if correlated with response studies to diet, will add further to possibilities for monitoring accurate prebiosis. In this respect, Langendijk *et al.* (1995) have reported the development of three 16S rRNA hybridization probes for the genus-specific detection of bifidobacteria in human faeces. Probes for other groups of gut micro-organisms are currently under development. However, these techniques should be subjected to rigorous evaluations of their fidelity and efficacy if they are to be applied on a quantitative basis.

A possible advantage that the probiotic approach has over prebiotics is that a wide variety of advantageous properties may be associated with the live additions ('optimal activity probiotics'). However, as mentioned earlier, survivability may be a problem. A further development is the use of synbiotics, where probiotics and prebiotics are combined. Thus, the live microbial additions would be used in conjunction with a specific substrate for growth. As such, improved survival and growth of the probiotic ought to occur. The feasibility of this concept has not yet been tested or validated, but it does offer increased efficacy to the probiotic approach. For example, the combination of a fructose-containing oligosaccharide with a *Bifidobacterium* strain is a potentially effective synbiotic, as is the use of lactitol in conjunction with lactobacilli. The concept could be taken further by the use of 'designer' synbiotics where the probiotic is genetically modified for improved metabolism of the prebiotic. Such advances are subject to the vagaries of legislation, but do at the very minimum offer interesting research tools.

With the advent of the 'functional food' concept, it is clear that there is an important niche for the pro-, pre- and synbiotic approaches. The implantation of live bacteria into the human, or selective increase of certain genera resident therein, is a functional claim. However, more rigorous research is required before health claims gain improved scientific credibility.

2.10 REFERENCES

Aguirre, M. and Collins, M.D. (1993) A review: lactic acid bacteria and human clinical infection. *J. Appl. Bacteriol.*, **75**, 95–107.

Alm, L. (1983) The effect of *Lactobacillus acidophilus* administration upon survival of *Salmonella* in randomly selected human carriers. *Prog. Food Nutr. Sci.*, **7**, 13–17.

Anand, S.K., Srinivasan, R.A. and Rao, L.K. (1985) Antibacterial activity associated with *Bifidobacterium bifidum*. *Cultured Dairy Prod. J.*, **2**, 21–3.

Andersson, R. (1986) Inhibition of *Staphylococcus aureus* and spheroplasts of Gram-negative bacteria by an antagonistic compound produced by *Lactobacillus plantarum*. *Int. J. Food Microbiol.*, **3**, 149–60.

Araya-Kojima, A., Yaeshima, T., Ishibashi, N. *et al.* (1995) Inhibitory effects of *Bifidobacterium longum* BB536 on harmful intestinal bacteria. *Bifid. Microflora*, **14**, 59–66.

Aronsson, B., Barany, P. and Nord, C.E. (1987) *Clostridium difficile* associated diarrhoea in uremic patients. *Eur. J. Clin. Microbiol.*, **6**, 352–6.

Axelsson, L.T., Chung, T.C., Dobrogosz, W.J. and Lindgren, S. (1989) Production of a broad spectrum antimicrobial substance by *Lactobacillus reuteri*. *Microb. Ecol. Health Dis.*, **2**, 131–6.

Barefoot, S.F., Nettles, C.G. and Chen, Y.R. (1994) Lactacin B, a bacteriocin produced by *Lactobacillus acidophilus*, in *Bacteriocins of Lactic Acid Bacteria* (eds L. De Vuyst and E.J. Vandamme), Blackie Academic and Professional, Glasgow, pp. 353–76.

Bartram, H-P., Scheppach, W., Gerlach, S. *et al.* (1994) Does yogurt enriched with *Bifidobacterium longum* affect colonic microbiology and fecal metabolites in healthy subjects? *Am. J. Clin. Nutr.*, **59**, 428–32.

Beck, C. and Necheles, H. (1961) Beneficial effects of administration of *Lactobacillus acidophilus* in diarrheal and other intestinal infections. *Am. J. Gastroenterol.*, **35**, 522–32.

Benno, Y. and Mitsuoka, T. (1992) Impact of *Bifidobacterium longum* on human fecal microflora. *Microbiol. Immunol.*, **36**, 683–94.

Benno, Y., Sawada, K. and Mitsuoka, T. (1984) The intestinal microflora of infants: composition of fecal flora in breast-fed and bottle-fed infants. *Microbiol. Immunol.*, **28**, 975–86.

Bernet, M.-F., Brassart, D., Neeser, J-R. and Servin, A.L. (1993) Adhesion of human bifidobacterial strains to cultured human intestinal epithelial cells and inhibition of enteropathogen–cell interactions. *Appl. Environ. Microbiol.*, **59**, 4121–8.

Black, F.T., Andersen, P.L., Arskov, J. *et al.* (1989) Prophylactic efficacy of lactobacilli on travellers diarrhoea. *Travel Med.*, **5**, 333–5.

Blomberg, L., Hentiksson, A. and Conway, P.L. (1992) Inhibition of adhesion of *Escherichia coli* K88 to piglet ileal mucus by *Lactobacillus spp. Appl. Environ. Microbiol.*, **59**, 34–9.

Bowden, T.A. Mansberger, A.R. and Lykins, L.E. (1981) Pseudomembranous enterocolitis: mechanism of restoring floral homeostasis. *Am. Surg.*, **47**, 173–83.

Braun, O.H. (1981) Effect of consumption of human milk and other formulas on intestinal bacterial flora in infants, in *Gastroenterology and Nutrition in Infancy* (ed. E. Lebenthal), Raven Press, New York, pp. 247–51.

Bullen, C.L., Teale, P.V. and Stewart, M.G. (1977) The effect of 'humanized' milks and supplemented breast feeding on the faecal flora of infants. *J. Med. Microbiol.*, **10**, 403–13.

Camarri, E., Belvisi, G., Guidoni, G. *et al.* (1981) A double blind comparison of two different treatments for acute enteritis in adults. *Chemotherapy*, **27**, 466–70.

Chan, R.C.Y., Reid, G., Irvin, R.T. *et al.* (1985) Competitive exclusion of uropathogens from human uroepithelial cells by *Lactobacillus* whole cells and cell fragments. *Infect. Immun.*, **47**, 84–9.

Chauviere, G., Coconnier, M.H., Kerneis, S. *et al.* (1992a) Competitive exclusion of diarrheagenic *Escherichia coli* (ETEC) from enterocyte-like Caco-2 cells in culture. *FEMS Microbiol. Letts*, **49**, 213–18.

Chauviere, G., Coconnier, M.H., Kerneis, S. *et al.* (1992b) Adherence of human *Lactobacillus acidophilus* onto human enterocyte-like cells, Caco-2 and HT-29 in culture. *J. Gen. Microbiol.*, **138**, 1689–96.

Clements, M.L., Levine, M.M., Black, R.E. *et al.* (1981) *Lactobacillus* prophylaxis for diarrhea due to enterotoxigenic *Escherichia coli*. *Antimicrob. Agents Chemother.*, **20**, 104–8.

Cohen, M.B. and Giannella, R.A. (1991) Bacterial infections: pathophysiology, clinical features and treatment, in *The Large Intestine: Physiology, Pathophysiology and Disease* (eds S.F. Phillips, J.H. Pemberton and R.G. Shorter), Raven Press, New York, pp. 395–428.

Cohen, P.S., Wadoloski, E.A. and Laux, D.C. (1986) Adhesion of human fecal *Escherichia coli* strain to a 50.5 kDal glycoprotein receptor present in mouse colonic mucus. *Microecol. Ther.*, **16**, 231–241.

Conway, P.L. and Henriksson, A. (1994) Strategies for the isolation and characterisation of functional probiotics, in *Human Health: The Contribution of Microorganisms* (ed. S.A.W. Gibson), Springer, London, pp. 75–93.

Conway, P.L. Gorbach, S.L. and Goldin, B.R. (1987) Survival of lactic acid bacteria in the human stomach and adhesion to intestinal cells. *J. Dairy Sci.*, **70**, 1–12.

Crociani, J., Grill, J.-P., Huppert, M. and Ballongue, J. (1995) Adhesion of different bifidobacteria strains to human enterocyte-like Caco-2 cells and comparison with *in vivo* study. *Lett. Appl. Microbiol.*, **21**, 146–8.

D'Appuzzo, V. and Salzberg, R. (1982) Die Behandlung der akuten Diarrhoe in der Padiatrie mit *Streptococcus faecium*: Resultate einer Doppleblindstudie. *Ther. Umsch.*, **39**, 1033–5.

De Simone, C., Ciardi, A., Grassi, A. *et al.* (1992) Effect of *Bifidobacterium bifidum* and *Lactobacillus acidophilus* on gut mucosa and peripheral blood B lymphocytes. *Immunopharmacol. Immunotoxicol.*, **14**, 331–40.

De Vuyst, L. and Vandamme, E.J. (1994) Antimicrobial potential of lactic acid bacteria, in *Bacteriocins of Lactic Acid Bacteria* (eds L. De Vuyst and E.J. Vandamme), Blackie Academic and Professional, Glasgow, pp. 91–142.

Dodd, H.M. and Gasson, M.J. (1994) Bacteriocins of lactic acid bacteria, in *Genetics and Biotechnology of Lactic Acid Bacteria* (eds M.J. Gasson and W.M. de Vos), Blackie Academic and Professional, Glasgow, pp. 211–51.

Drasar, B.S. and Roberts, A.K. (1990) Control of the large bowel microflora, in *Human Microbial Ecology* (eds M.J. Hill and P.D. Marsh), CRC Press, Boca Raton, FL, pp. 87–111.

Duffy, L.C., Zielezny, M.A., Riepenhoff-Talty, M. *et al.* (1994a) Effectiveness of *Bifidobacterium bifidum* in mediating the clinical course of murine rotavirus diarrhea. *Ped. Res.*, **35**, 690–5.

Duffy, L.C. Zielezny, M.A., Riepenhoff-Talty, M. *et al.* (1994b) Reduction of virus shedding by *B. bifidum* in experimentally induced MRV infection. *Dig. Dis. Sci.*, **39**, 2334–40.

Faure, J.C., Schellenberg, D.A., Bexter, A. and Wuerzuer, H.P. (1984) Barrier effect of *Bifidobacterium longum* on a pathogenic *Escherichia coli* strain by gut colonization in the germ-free rat. *Z. Ernahrung.*, **23**, 41–4.

Finegold, S.M., Sutter, V.L. and Mathisen, G.E. (1983) Normal indigenous intestinal flora, in *Human Intestinal Microflora in Health and Disease* (ed. D.J. Hentges), Academic Press, London, pp. 3–31.

Freter, R. (1992) Factors affecting the microecology of the gut, in *Probiotics: The Scientific Basis* (ed. R. Fuller), Chapman & Hall, London, pp. 111–44.

Friend, A., Farmer, R.E. and Shahani, K.M. (1982) Effect of feeding and intraperitoneal implantation of yoghurt culture cells on Ehrlich ascites tumor. *Milchwissenschaft*, **37**, 708–10.

Fuller, R. (1989) A review: probiotics in man and animals. *J. Appl. Bacteriol.*, **66**, 365–78.

Fuller, R. (ed.) (1992) *Probiotics: The Scientific Basis*, Chapman & Hall, London.

Fuller, R. (1994) Probiotics: an overview, in *Human Health: The Contribution of Microorganisms* (ed. S.A.W. Gibson), Springer, London, pp. 61–73.

Gauhe, A.P. Gyorgy, P.A. Hoover, J.R.E. *et al.* (1954) Bifidus factor. Preparations obtained from human milk. *Arch. Biochem.*, **48**, 214–24.

Geis, A. (1989) Antagonistic compounds produced by lactic acid bacteria. *Kiel Milchwirt Forschung*, **41**, 97–104.

Gibson, G.R. and MacFarlane, G.T. (eds) (1995) *Human Colonic Bacteria. Role in Physiology, Pathology and Nutrition*, CRC Press, Boca Raton, FL.

Gibson, G.R. and Roberfroid, M.B. (1995) Dietary modulation of the human colonic microbiota: introducing the concept of prebiotics. *J. Nutr.*, **125**, 1401–12.

Gibson, G.R. and Wang, X. (1994) Regulatory effects of bifidobacteria on other colonic bacteria. *J. Appl. Bacteriol.*, **77**, 412–20.

Gibson, G.R., Beatty, E.B., Wang, X. and Cummings, J.H. (1995) Selective stimulation of bifidobacteria in the human colon by oligofructose and inulin. *Gastroenterology*, **108**, 975–82.

Gilliland, S.E. and Kim, H.S. (1982) Effect of viable starter culture in yogurt on lactose utilization in humans. *J. Dairy Sci.*, **67**, 1–6.

Gilliland, S.E. and Speck, M.L. (1977) Deconjugation of bile acids by intestinal lactobacilli. *Appl. Environ. Microbiol.*, **33**, 15–18.

Gilliland, S.E., Nelson, C.R. and Maxwell, C. (1985) Assimilation of cholesterol by *Lactobacillus acidophilus*. *Appl. Environ. Microbiol.*, **49**, 377–81.

Gorbach, S.L., Chang, T. and Goldin, B. (1987) Successful treatment of relapsing *Clostridium difficile* colitis with *Lactobacillus* GG. *Lancet*, ii, 1519.

Gorbach, S.L., Barza, M., Giuliano, M. and Jacobus, N.V. (1988) Colonization resistance of the human intestinal microflora: testing the hypothesis in normal volunteers. *Eur. J. Clin. Microbiol. Infect. Dis.*, **7**, 98–102.

Graf, W. (1983) Studies on the therapeutic properties of acidophilus milk, in *Nutrition and the Intestinal Flora, Symposia of the Swedish Nutrition Foundation XV* (ed. B. Halgren), pp. 119–21.

Gyorgy, P.A. (1953) Hitherto unrecognized biochemical differences between human milk and cow's milk. *J. Ped.*, **11**, 98–108.

Halpern, G.M., Vruwing, K.G., Van de Water, J. *et al.* (1991) Influence of long-term yogurt consumption in young adults. *Int. J. Immunother.*, **VII**, 205–10.

Hamilton-Miller, J.M.T., Shah, S. and Smith, C.T. (1996) 'Probiotic' remedies are not what they seem. *Br. Med. J.*, **312**, 55–6.

Hayakawa, K., Mizutani, J., Wada, K. *et al.* (1990) Effects of soybean oligosaccharides on human faecal microflora. *Microbiol. Ecol. Health Dis.*, **3**, 292–303.

Hentges, D.J. (1983) Role of the intestinal microflora in host defense against infection, in *Human Intestinal Microflora in Health and Disease* (ed. D.J. Hentges), Academic Press, London, pp. 311–32.

Hidaka, H., Eida, T., Takiwaza, T. *et al.* (1986) Effects of fructooligosaccharides on intestinal flora and human health. *Bifid. Microflora*, **5**, 37–50.

Hill, M.J. (1986) Factors affecting bacterial metabolism, in *Microbial Metabolism in the Digestive Tract* (ed. M.J. Hill), CRC Press, Boca Raton, FL, pp. 22–8.

Hotta, M., Sato, Y., Iwata, S. *et al.* (1987) Clinical effects of *Bifidobacterium* preparations on pediatric intractable diarrhea. *Keio J. Med.*, **36**, 298–314.

Huis in't Veld, J.H.J. and Havenaar, R. (1993) Selection criteria for microorganisms for probiotic use, in *Probiotics and Pathogenicity, Flair No. 6* (eds. J.F. Jensen, M.H. Hinton and R.W.A.W. Mulder), DLO Spelderholt Centre for Poultry Research and Information Services, pp. 11–19.

Isolauri, E., Juntunen, M., Rautanen, T. *et al.* (1991) A human *Lactobacillus* strain (*Lactobacillus casei* sp. strain GG) promotes recovery from acute diarrhea in children. *Pediatrics*, **88**, 90–7.

Isolauri, E., Joensuu, J., Suomalainen, H. *et al.* (1995) Improved immunogenicity of oral D × RRV reassortant rotavirus vaccine by *Lactobacillus casei* GG. *Vaccine* **13**, 310–12.

Itoh, T., Fujimoto, Y., Kawai, Y. *et al.* (1995) Inhibition of food-borne pathogenic bacteria by bacteriocins from *Lactobacillus gasseri. Lett. Appl. Microbiol.*, **21**, 137–41.

Johansson, M.-L., Molin, G., Jeppsson, B. *et al.* (1993) Administration of different *Lactobacillus* strains in fermented oatmeal soup: *in vivo* colonization of human intestinal mucosa and effect on the indigenous flora. *Appl. Environ. Microbiol.*, **59**, 15.

Kaila, M., Isolauri, E., Soppi, E. *et al.* (1992) Enhancement of the circulating antibody secreting-cell response on human diarrhea by a human *Lactobacillus* strain. *Pediatr. Res.*, **32**, 141–4.

Kaloud, H. and Stogmann, W. (1969) Clinical experience with a bifidus milk feed. *Arch. Kinderheilk*, **177**, 29–35.

Kawai, Y., Saito, T., Toba, T. *et al.* (1994) Isolation and characterization of a highly hydrophobic new bacteriocin (Gassericin A) from *Lactobacillus gasseri* LA 39. *Biosci. Biotechnol. Biochem.*, **58**, 1218–21.

Kitazawa, H., Tomioka, Y., Matsdumura, K. *et al.* (1994) Expression of mRNA encoding IFNα in macrophages stimulated with *Lactobacillus gasseri. FEMS Microbiol. Letts*, **120**, 315–22.

Klaenhammer, T.R., Ahn, C., Fremaux, C. and Milton, K. (1992) Molecular properties of *Lactobacillus* bacteriocins, in *Bacteriocins, Microcins and Lantibiotics* (eds. R. James, C. Ladzunski and F. Pattus), NATA Series H, 65, Springer, Berlin, pp. 37–58.

Klaenhammer, T.R. Ahn, C. and Muriana, P.M. (1994) Lactacin F, a small hydrophobic heat-stable bacteriocin from *Lactobacillus johnsonii*, in *Bacteriocins of Lactic Acid Bacteria* (eds. L. De Vuyst and E.J. Vandamme), Blackie Academic and Professional, Glasgow, pp. 377–96.

Kleeman, E.G. and Klaenhammer, T.R. (1982) Adherence of *Lactobacillus* species to human fetal intestinal cells. *J. Dairy Sci.*, **65**, 2065–9.

Kohwi, Y., Imai, K., Tamura, Z. and Hasimoto, Y. (1978) Antitumor effect of *Bifidobacterium infantis* in mice. *Gann*, **69**, 613–18.

Kohwi, Y., Hashimoto, Y. and Tamura, Z. (1982) Antitumor and immunological adjuvant effect of *Bifidobacterium infantis* in mice. *Bifid. Microflora*, **1**, 61–4.

Langendijk, P.S., Schut, F., Jansen, G.J., Raangs, G.C., Kamphuis, G.R., Wilkinson, M.H.F. and Welling, G.W. (1995) Quantitative fluorescence in situ hybridization of *Bifidobacterium* spp. with genus-specific 16S rRNA-targeted probes and its application in fecal samples. *Appl. Environ. Microbiol.*, **61**, 3069–75.

Lee, Y.K. and Salminen, S. (1995) The coming age of probiotics. *Trend Food Sci. Technol.*, **6**, 241–5.

Lidbeck, A., Gustafsson, J.-A. and Nord, C.E. (1987) Impact of *Lactobacillus acidophilus* supplements on the human oropharyngeal and intestinal microflora. *Scand. J. Infect. Dis.*, **19**, 531–7.

Lin, S.Y., Ayres, J.W., Winkler, W. and Sandine, W.E. (1989) *Lactobacillus* effects on cholesterol: *in vitro* and *in vivo* results. *J. Dairy Sci.*, **72**, 2885–99.

Link-Amster, H., Rochat, F., Saudan, K.Y. _et al._ (1994) Modulation of a specific humoral immune response and changes in intestinal flora mediated through fermented milk intake. _FEMS Immunol. Med. Microbiol._, **10**, 55–64.

MacFarlane, G.T. and Gibson, G.R. (1995) Bacterial infections and diarrhea, in _Human Colonic Bacteria: Role in Nutrition, Physiology and Pathology_ (eds G.R. Gibson and G.T. MacFarlane), CRC Press, Boca Raton, FL, pp. 201–26.

Mantere-Alhonen, S., Noro, K. and Sippola, L. (1989) Vorlaufige untersuchugen uber den antibackteriellen einfluss einiger bifidobakterien-und Lacktobazillen-arten. _Fin. J. Dairy Sci._, **XLVII**, 19–28.

Mayer, J.B. (1969) Interrelationships between diet, intestinal flora and viruses. _Phyik. Med. Rehab._, **10**, 16–23.

McGroaty, J.A., Hawthorn, A.A. and Reid, G. (1988) Anti-tumour activity of lactobacilli _in vitro. Microbios Lett._, **39**, 105–12.

Meghrous, J., Euloge, P., Junelles, A.M. _et al._ (1990) Screening of _Bifidobacterium_ strains for bacteriocin production. _Biotechnol. Lett._, **12**, 575–80.

Mehta, A.M., Patel, K.A. and Dave. P.J. (1983) Isolation and purification of an inhibitory protein from _Lactobacillus acidophilus. Microbios_, **37**, 37–43.

Mitsuoka, T. (1984) Taxonomy and ecology of bifidobacteria. _Bifid. Microflora_, **3**, 11–28.

Mitsuoka T., Hidaka, H. and Eida, T. (1987) Effect of fructooligosaccharides on intestinal microflora. _Die Nahrung_, **31**, 426–36.

Modler, H.W., McKellar, R.C. and Yaguchi, M. (1990) Bifidobacteria and bifidogenic factors. _Can. Inst. Food Sci. Technol._, **23**, 29–41.

Moore, W.E.C. and Holdeman, L.V. (1974) Human fecal flora. The normal flora of 20 Japanese Hawaiians. _Appl. Microbiol._, **27**, 961–79.

Moore, W.E.C. and Moore, L.H. (1995) Intestinal floras of populations that have a high risk of colon cancer. _Appl. Environ. Microbiol._, **61**, 3202–7.

Nes, I.F., Mortvedt, C.I., Nissen-Meyer, J. and Skaugen, M. (1994) Lactocin S, a lanthionine-containing bacteriocin isolated from _Lactobacillus sake_ L45, in _Bacteriocins of Lactic Acid Bacteria_ (eds L. De Vuyst and E.J. Vandamme), Blackie Academic and Professional, Glasgow, pp. 435–49.

Oksanen, P.J., Salminen, S., Saxelin, M. _et al._ (1990) Prevention of traveller's diarrhoea by _Lactobacillus_ GG. _Ann. Med._, **22**, 53–6.

O'Riordan, K.C., Condon, S. and Fitzgerald, G.F. (1995) Bacterial interference by _Bifidobacterium_ species and a comparative analysis of genomic profiles from strains of this genus. _Proc. Lactic Acid Bacteria Conference, Cork, Ireland_, p.207.

Op den Camp, H.J.M., Oosterhoh, A and Veerkamp, J.H. (1985) Interaction of bifidobacterial lipoteichoic acid with human intestinal epithelial cells, _Infect. Immun._, **47**, 332–4.

Ouwenhand, A.C. and Conway, P.L. (1996) Purification and characterization of a component produced by _Lactobacillus fermentum_ that inhibits the adhesion of K88 expressing _Escherichia coli_ to porcine ileal mucus. _J. Appl. Bacteriol._, **80**, 311–18.

Pearce, J.L. and Hamilton, J.R. (1974) Controlled trial of orally administered lactobacilli in acute infantile diarrhoea. _J. Ped._, **84**, 261–2.

Perdigon, G. and Alvarez, S. (1992) Probiotics and the immune state, in _Probiotics: The Scientific Basis_ (ed. R. Fuller), Chapman & Hall, London, pp. 146–80.

Perdigon, G., Medici, M., Bibas Bonet de Jorrat, M.E. _et al._ (1993) Immunomodulating effects of lactic acid bacteria on mucosal and tumoral immunity. _Int. J. Immunother._, **IX**, 29–52.

Perdigon, G., Rachid, M., De Budeguer, M.V. and Valdez, J.C. (1994) Effects of yoghurt feeding on the small and large intestine associated lymphoid cells in mice. _J. Dairy Res._, **61**, 553–62.

Perdigon, G., Aguero, G., Alvarez, S. *et al.* (1995) Effect of viable *Lactobacillus casei* feeding on the immunity of the mucosae and intestinal microflora in malnourished mice. *Milchwiss*, **50**, 251–6.

Quigley, M.E. and Kelly, S.M. (1995) Structure, function and metabolism of host mucus glycoproteins, in *Human Colonic Bacteria: Role in Nutrition, Physiology and Pathology* (eds G.R. Gibson and G.T. MacFarlane), CRC Press, Boca Raton, FL, pp. 175–99.

Rammelsberg, M. and Radler, F. (1990) Antibacterial polypeptides of *Lactobacillus* species. *J. Appl. Bacteriol.*, **69**, 177–84.

Rastegar Lari, A., Gold, L.F., Borderon, J.C. *et al.* (1990) Implantation and *in vivo* antagonistic effects of antibiotic susceptible *Escherichia coli* strains administered to premature newborns. *Biol. Neonat.*, **56**, 73–8.

Reddy, G.V., Friend, B.A., Shahani, K.M. and Farmer, R.E. (1983) Antitumor activity of yogurt components. *J. Food Protect.*, **46**, 8–11.

Reid, G., Servin, A., Bruce, A.W. and Busscher, H.J. (1993) Adhesion of three *Lactobacillus* strains to human urinary and intestinal epithelial cells. *Microbios*, **75**, 57–65.

Robinson, E.L. and Thompson, W.L. (1952) Effect on weight gain of the addition of *Lactobacillus acidophilus* to the formula of newborn infants. *J. Ped.*, **41**, 395–8.

Rolfe, R.D., Helebian, S. and Finegold, S.M. (1981) Bacterial interference between *Clostridium difficile* and normal fecal flora. *J. Infect. Dis.*, **143**, 470–5.

Rowland, I.R. and Tanaka, R. (1993) The effects of transgalactosylated oligosaccharides on gut flora metabolism in rats associated with a human faecal microflora. *J. Appl. Bacteriol.*, **74**, 667–74.

Rumney, C. and Rowland, I.R. (1995) Non-digestible oligosaccharides – potential anti-cancer agents? *BNF Nutr. Bull.*, **20**, 194–203.

Saavedra, J.M. (1995) Microbes to fight microbes: a not so novel approach for controlling diarrheal disease. *J. Ped. Gastroenterol. Nutr.*, **21**, 125–9.

Saavedra, J.M. Bauman, N.A., Oung, I. *et al.* (1994) Feeding of *Bifidobacterium bifidum* and *Streptococcus thermophilus* to infants in hospital for prevention of diarrhoea and shedding of rotavirus. *Lancet*, **344**, 1046–9.

Salminen, S., Isolauri, E. and Onnela, T. (1995) Gut flora in normal and disordered states. *Chemotherapy*, **41**, 5–15.

Sato, J., Mochizuki, K. and Homma, N. (1982) Affinity of the *Bifidobacterium* to intestinal mucosal epithelial cells. *Bifid. Microflora*, **1**, 51–4.

Savage, D.C. and Fletcher, M. (1985) *Bacterial Adhesion: Mechanisms and Physiological Significance*, Plenum Press, New York.

Scardovi, V. (1986) Genus *Bifidobacterium*, in *Bergey's Manual of Systematic Bacteriology*, Vol. 2 (ed. N.S. Mair), Williams & Wilkins, New York, pp. 1418–34.

Schillinger, U. (1994) Sakacin A produced by *Lactobacillus sake* Lb 706, in *Bacteriocins of Lactic Acid Bacteria* (eds L. De Vuyst and E.J. Vandamme), Blackie Academic and Professional, Glasgow, pp. 419–34.

Schwan, A., Sjolin, S., Trottestam, U. and Aronsson, B. (1984) Relapsing *Clostridium difficile* enterocolitis cured by rectal infusion of normal faeces. *Scand. J. Infect. Dis.*, **16**, 211–15.

Sears, H.J., Janes, H., Saloum, R. *et al.* (1956) Persistence of individual strains of *Escherichia coli* in man and dog under varying conditions. *J. Bacteriol.*, **71**, 370–2.

Sekine, K., Toida, T., Saito, M. *et al.* (1985) A new morphologically characterized cell wall preparation (whole peptidoglycan) from *Bifidobacterium infantis* with a higher efficacy on the regression of an established tumor in mice. *Cancer Res.*, **45**, 1300–7.

Shinefield, H.R., Ribble, J.C. and Boris, M. (1971) Bacterial interference between strains of *Staphylococcus aureus*. *Amr. J. Dis. Child.*, **121**, 148–52.

Silva, M., Jacobus, N.V., Deneke, C. and Gorbach, S.L. (1987) Antimicrobial substance from a human *Lactobacillus* strain. *Antimicrob. Agents Chemother.*, **31**, 1231–3.

Smith, H. (1995) The revival of interest in mechanisms of bacterial pathogenicity. *Biol. Rev.*, **70**, 277–316.

Solis, P.B. and Lemmonnier, D. (1993) Induction of human cytokines by bacteria used in dairy foods. *Nutr. Res.*, **13**, 1127–40.

Surawicz, C.M., Elmer, G., Speelman, P. *et al.* (1989a) Prevention of antibiotic-associated diarrhea by *Saccharomyces boulardii*: a prospective study. *Gastro-enterology*, **96**, 981–8.

Surawicz, C.M., McFarland, L.V., Elmer, G. and Chinn, J. (1989b) Treatment of recurrent *Clostridium difficile* colitis with vancomycin and *Saccharomyces boulardii*. *Am. J. Gastroenterol.*, **84**, 1285–7.

Tamura, Z. (1983) Nutriology of bifidobacteria. *Bifid. Microflora*, **2**, 3–16.

Tannock, G.W. (1995) The role of probiotics, in *Human Colonic Bacteria: Role in Nutrition, Physiology, and Pathology* (eds G.R. Gibson and G.T. MacFarlane), CRC Press, Boca Raton, FL, pp. 257–71.

Tasvac, B. (1964) Infantile bacterial enterocolitis. Treatment with *Bifidobacterium bifidum*. *Ann. Ped.*, **11**, 291–7.

Thompson, J.K., Collins, M.A. and Mercer, W.D. (1996) Characterization of a proteinaceous antimicrobial produced by *Lactobacillus helveticus* CNRZ450. *J. Appl. Bacteriol.*, **80**, 338–48.

Tichaczek, P.S., Nissen-Meyer, J.I.F., Vogel, R.F. and Hammes, W.P. (1992) Characterization of the bacteriocins curvacin A from *Lactobacillus curvatus* LTH1174 and sakacin P from *L. sake* LTH673. *Sys. Appl. Microbiol.*, **15**, 460–80.

Tojo, M., Oikawa, T., Morikawa, Y. *et al.* (1987) The effects of *Bifidobacterium breve* administration on *Campylobacter* enteritis. *Acta Ped. Jpn.*, **29**, 160–7.

Tomic-Karovic, K. and Fanjek, J.J. (1962) Acidophilus milk in therapy of infantile diarrhea caused by pathogenic *Escherichia coli*. *Ann. Pediatr.*, **199**, 625–34.

Tvede, M. and Rask-Madsen, J. (1989) Bacteriotherapy for chronic relapsing *Clostridium difficile* diarrhoea in six patients. *Lancet*, **I**, 1156–60.

Upreti, G.C. and Hindsdill, R.D. (1975) Production and mode of action of lactocin 27: bacteriocin from a homofermentative *Lactobacillus*. *Antimicrob. Agents Chemother.*, **7**, 139–45.

Van Loo, J., Coussement, P., De Leenheer, L. *et al.* (1995) On the presence of inulin and oligofructose as natural ingredients in the Western diet. *CRC Rev. Food Sci. Technol.*, **35**, 525–52.

Wadstrom, T., Andersson, K., Sydow, M. *et al.* (1987) Surface properties of lacto-bacilli isolated from the small intestine of pigs. *J. Appl. Bacteriol.*, **62**, 513–20.

Wang, X. and Gibson, G.R. (1993) Effects of *in vitro* fermentation of oligofruc-tose and inulin by bacteria growing in the human large intestine. *J. Appl. Bacteriol.*, **75**, 373–80.

Willis, A.T., Bullen, C.L., Williams, K. *et al.* (1973) Breast milk substitute: a bacteriological study. *Br. Med. J.*, **4**, 67.

Wilson, K.H. and Perini, F. (1988) Role of competition for nutrients in suppression of *Clostridium difficile* by the colonic microflora. *Infect. Immun.*, **56**, 2610–14.

Wood, B.J.B. (ed.) (1992) *The Lactic Acid Bacteria in Health and Disease*, Elsevier Applied Science, Barking, Essex.

Wolin, M.J. (1974) Metabolic interaction among intestinal microorganisms. *Am. J. Clin. Nutr.*, **27**, 1320–24.

Yasui, H. and Ohwaki, M. (1991) Enhancement of immune response in Peyer's patch cells cultured with *Bifidobacterium breve. J. Dairy Sci.*, **74**, 1187–95.

Yasui, H., Kiyoshima, J. and Ushima, H. (1995) Passive protection against rotavirus-induced diarrhea of mouse pups born to and nursed by dams fed *Bifidobacterium breve* YIT4064. *J. Infect. Dis.*, **172**, 403–9.

Yasui, H., Nagaoka, A., Mike, A. *et al.* (1992) Detection of *Bifidobacterium* strains that induce large quantities of IgA. *Microbiol. Ecol. Health Dis.*, **5**, 155–62.

Yawaza, K. and Tamura, Z. (1982) Search for sugar sources for selective increase of bifidobacteria. *Bifid. Microflora*, **1**, 39–44.

Yawaza, K., Imai, K. and Tamura, Z. (1978) Oligosaccharides and polysaccharides specifically utilisable by bifidobacteria. *Chem. Pharm. Bull.*, **26**, 3306–11.

Yoshita, M., Fujita, K., Sakata, H. *et al.* (1991) Development of the normal intestinal flora and its clinical significance in infants and children. *Bifid. Microflora*, **10**, 11–27.

Zychowicz, C., Surazynska, A., Siewierska, B. and Cieplinska, T. (1974) Effect of *Lactobacillus acidophilus* culture (acidophilus milk) on the carrier state of *Shigella* and *Salmonella* organisms in children. *Ped. Polska.*, **49**, 997–1003.

CHAPTER 3

Antibiotic-associated diarrhoea: treatments by living organisms given by the oral route (probiotics)

G. Corthier

3.1 INTRODUCTION

Pseudomembranous colitis and a large percentage (about 30%) of antibiotic-associated diarrhoea are caused by the overgrowth of *Clostridium difficile* in the human digestive tract. This overgrowth is mainly due to disorders in gut bacterial populations leading to disruption of the colonization resistance to pathogens (Hentges, 1992). *C. difficile* produces two toxins (an enterotoxin and a cytotoxin) which are responsible for the disease (Figure 3.1). Medical treatments mostly involve antibiotics (vancomycin or metronidazole) that do not facilitate the restoration of the colonization resistance, so that relapses are often observed. The aim of this chapter is to overview *C. difficile*-associated enteropathies as well as bacterial toxins and to present the current and future applications of probiotic treatments in a preventive or curative mode.

3.2 ANTIBIOTIC-ASSOCIATED DIARRHOEA AND PSEUDOMEMBRANOUS COLITIS

This subject has been reviewed by different authors (among them Bartlett, 1979, 1994; Lyerly, Krivian and Wilkins, 1988; and Rolfe and

Probiotics 2: Applications and practical aspects.
Edited by R. Fuller.
Published in 1997 by Chapman & Hall. ISBN 0 412 73610 1

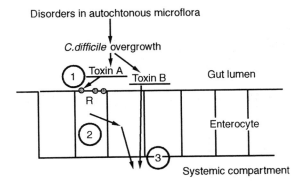

Fig. 3.1 General outline of pathology due to *Clostridium difficile*. 1, attachment to brush border receptor (R); 2, enterotoxicity and cell death; 3, inflammation process and dissemination of both toxins (A and B) in systemic circulation.

Finegold, 1988, in a book containing 16 reviews on the subject). In order to reduce the number of cited references, only the studies not mentioned in these reviews will be cited here.

At any age of life composition of the intestinal microflora represents a fragile equilibrium between different bacterial populations which can be predominant, subdominant or transitory. The global functions of the microflora depend on this quantitative and qualitative equilibrium. Antibiotic treatments induce a complete or partial destruction of the intestinal microflora. At the end of the treatment, bacteria hidden in niches, in the environment or provided by food colonize the gut. However, the initial equilibrium involved in crucial functions such as colonization resistance to pathogen bacteria is not easily restored (Hentges, 1992).

Antibiotic treatments often lead to antibiotic-associated diarrhoea, the extreme form of which is pseudomembranous colitis. Antibiotic-associated colitis was recognized (Bartlett, 1979) as 'an intriguing paradox of medical progress that compromises the therapeutic utility of an important group of antibiotics, adds a potential obstacle to new drug development and challenges our ability to fashion the welter of clinical and pathologic observations into a cohesive disease entity'.

3.2.1 The pathological effects

An antibiotic-associated diarrhoea generally occurs following several days of antibiotic treatment but it can also be observed at the end of the treatment or during the following 6 weeks. The severity of diarrhoea can range from a mild increase in stool frequency, sometimes associated with an abdominal pain and fever, to debilitating diarrhoea.

Table 3.1 Causative agents involved in *Clostridium difficile* pseudomembranous colitis or diarrhoea. From Rolfe and Finegold (1988)

General classification	Specific agent
Penicillins	Methicillin, nafcillin, oxacillin, cloxacillin, dicloxacilin Penicillin G, penicillin V Ampicillin Amoxycillin Carbenicillin, ticarcillin
First-generation cephalosporins	Cephalothin Cefazolin Cephradine Cephalexin
Second-generation cephalosporins	Cefamandole Cefoxitin
Third-generation cephalosporins and miscellaneous beta-lactams	Moxalactam Cefotaxime Ceftizoxime Cefoperazone Imipenen
Lincosamides	Lincomycin Clindamycin
Macrolides/tetracyclines	Erythromycin Spiramycin Tetracycline
Miscellaneous antimicrobials	Chloramphenicol Metronidazole Sulphisoxazole Sulphamethoxazole-trimethoprim Sulphasalazine Rifampicin Aminoglycosides (given orally) Amphotericin Miconazole
Cancer chemotherapeutics	Adriamycin Cyclophosphamide 5-Fluorouracil Methotrexate

Although usually self-limiting upon termination of the antibiotic, it may not always be feasible to stop antibiotherapy. In some patients diarrhoea may exacerbate an already serious condition by dehydration and/or perturbation of the electrolyte balance. A significant percentage (20–30%) of such diarrhoea is due to *C. difficile* overgrowth, some of them are attributed to *Klebsiella oxytoca* and others are probably due to metabolic disorders. The incriminated antibiotics have a wide spectrum. The most famous is clindamycin (for pseudomembranous colitis) but it has been shown that almost all antibiotic treatments could lead to antibiotic-associated diarrhoea or pseudomembranous colitis (Table 3.1). Pseudomembranous colitis is the most severe form and probably the last stage of a spectrum of acute intestinal diseases mainly due to antibiotic administration and associated with *in situ* proliferation of toxinogenic *C. difficile*. This pathogen also seems to be responsible for cases of pseudomembranous colitis not linked to antibiotherapy.

3.2.2 Involvement of *C. difficile*

Some cases of haemorragic colitis in the right colon have been attributed to *Klebsiella oxytoca* (Benoit *et al.*, 1992) which produces a cytotoxin (Minami *et al.*, 1992). They generally regress after termination of the antibiotic treatment.

Although pseudomembranous colitis was first described in 1893 by Finney, clear involvement of *C. difficile* was only established in the period 1975–78 (Bartlett, 1994). It was demonstrated that in diarrhoeal specimens from pseudomembranous colitis patients, the cytotoxic activity was not caused by viruses but was produced by some species of *Clostridia* since it was neutralized by antisera against *Clostridium sordellii*. Attempts to isolate *C. sordellii* from cytotoxic faecal specimens failed but it was shown that *C. difficile* could be consistently isolated from these stools. The strains caused a similar type of disease in antibiotic-treated hamsters.

C. difficile is a Gram-positive, sporulating anaerobic bacterium (usually mobile) of 0.5–1.6 by 3.0–16.9 µm (spore in subterminal position). It was first described by Hall and O'Toole in 1935 and named *Bacillus difficilis* because its isolation and culture were difficult. *C. difficile* produces two toxins (see below) which are responsible for the pathological effects. The different *C. difficile* isolates can be typed according to the molecular weight of proteins (sodium dodecyl sulphate polyacrylamide gel electrophoresis (SDS-PAGE), Tabaqchali group's classification) or antigenic properties (serogroups, Delmée group's classification). For instance, the D serogroup of *C. difficile* does not produce any toxin and has never been involved in pathologies while C, G and H serogroups have often been associated with nosocomial infections. These two classifications are useful for epidemiological studies.

3.2.3.   Toxins of *C. difficile*

Two toxins are produced by *C. difficile*. One of them, toxin A, possesses both an enterotoxic activity and a cytotoxic activity. The other one, toxin B, is an extremely potent cytotoxin. Toxin B is responsible for the cytotoxic activity observed in faecal specimens from persons with pseudomembranous colitis because it masks the cytotoxic activity of the toxin A. It is currently admitted that both toxins are implicated in *C. difficile* disease: experimental animals (hamsters) have to be vaccinated against both toxins to be protected against the disease. Pure toxins given orally to experimental animals mimicked the pathology observed in humans. Both toxins were consistently detected in faecal specimens from humans and experimental animals. As a result, much of the work on diseases caused by *C. difficile* focused on these toxins. Two toxins produced by *C. sordellii* have been described: the haemorrhagic toxin (HT) and the lethal toxin (LT), which are similar to toxin A and B of *C. difficile* and share common antigens (Popoff, 1987; Martinez and Wilkins, 1992). These observations explain the neutralization of *C. difficile* toxins by antisera to *C. sordellii*, which is currently used for diagnosis.

(a)   Genetic organization

The genes encoding toxins A and B (toxA and toxB) have been cloned and sequenced (Barroso *et al.*, 1984; Dove, Wang and Price, 1990). The two genes are located approximately 1 kb apart and are separated by a small open reading frame of 0.4 kb (unknown function). The toxA and toxB genes are 8.1 kb and 7.1 kb and encode polypeptides of 308 000 kd and 269 000 kd. Both toxins are produced as single polypeptides, making them the largest bacterial toxins known. The two toxins have a complex series of contiguous repeating units at the carboxyl-terminus comprising about one-third of each molecule. In toxin A, these units are the binding portion recognized by monoclonal antibodies (Corthier *et al.*, 1991). The toxins share common structural features supporting the idea that the toxA and toxB genes evolved by gene duplication (Eichel-Streiber *et al.*, 1992).

The promoters of both toxins have not been identified. Are there different promoters for each toxin or a unique one for the whole operon? It is very important to answer the question of toxin regulation because it will help to find new approaches of protection (especially in the probiotic field). The regulation is probably global for both toxins since they are produced in similar amounts in almost all *C. difficile* isolates. No strain was found to produce toxin A without toxin B.

Table 3.2 Main properties of toxin A and B

	Toxin A	Toxin B
Molecular weight	308 057 Da	269 709 Da
Isoelectric point	5.3	4.1
Number of amino acids	2710	2366
Number of repeated sequences in C terminal region	38	24
Cytotoxic activity on current cell lines	+(10 ng)	+++(5 fg)
Enterotoxic activity	+	−
Lethal activity	+	+
Haemagglutination	+	−

(b) Global mechanism of action

The general properties are currently known though some points have to be clarified (Table 3.2). Toxin A binds to its receptors along the gut mucosa and causes extensive damage, resulting in inflammation, necrosis and fluid in the intestinal lumen. Toxin B then exits the gut via the tissue damaged caused by toxin A. A neuronal involvement of toxin A in the intestinal effects was described (Castagliuolo *et al.*, 1994). Target receptors and cells outside the intestine have not been identified. However, both toxins may exert their toxic effect on cells located elsewhere than at the initial site of infection. Both toxins are lethal in similar conditions when inoculated systematically into animals.

This mechanism could be more complicated. A recent work (Riegler *et al.*, 1995) indicates that the human colon is about 10 times more sensitive to the damaging effects of toxin B than to those of toxin A, suggesting that toxin B plays a more important role than toxin A in the pathogenesis of *C. difficile* colitis in man. Adhesion of *C. difficile* (after heat shock treatment) to cell line has been demonstrated. The genetic determinant of this process has been cloned (Karjalainen *et al.*, 1994) but involvement of *C. difficile* attachment in human disease has not yet been clearly established.

(c) The toxins and their receptors

The main properties of toxin A and B are summarized in Table 3.2. The first step of the pathology is the fixation of toxin A to the brush-border receptor. Such a molecule was identified in rabbit red blood cells as a trisaccharid: Galα1-3Galβ1-4GlcNAc. However, it was not expressed in intestinal cells from rabbits or humans. In rabbit enterocytes, a galactose-containing glycoprotein, N-acetylglucosamine, coupled to G protein, was identified as receptor (Pothoulakis *et al.*, 1991). It was reduced in newborn rabbits which were not sensitive to the toxin (Eglow *et al.*, 1992). In human enterocytes, the receptor to toxin A has not been identified. When toxin A is internalized, it sticks to actin and disorganizes the cytoskeleton. Toxin A is sensitive to intestinal protease but it induces a release of the tight junction between enterocytes (Hecht *et al.*, 1988) leading to serum outcome in the digestive lumen which neutralizes proteases (Corthier *et al.*, 1989), increasing the pathological process.

The cellular receptor to toxin B has not been identified but, as toxin B is active on numerous cell lines, its receptor is probably ubiquitous. After toxin internalization, a depolymerization of actin filament is observed leading to cell rounding and cytotoxicity. The first effect of toxin B is probably mediated by a modification of the Rho protein involved in actin synthesis regulation (Just *et al.*, 1995)

3.2.4   *C. difficile* in infant digestive tract

Human babies very frequently harbour high levels of *C. difficile* in their faeces but *C. difficile*-related diseases are rare. The gastrointestinal tract of healthy newborns is frequently (15–63%) colonized by *C. difficile* during the first 2 weeks of life. Then the strain remains established until the age of 1–2 years. At that age, the prevalence ranges from 7 to 48%. Strains isolated from infants have often proven to be highly toxinogenic *in vitro*, resembling those isolated from stools of adults suffering from pseudomembranous colitis. Many asymptomatic infants colonized with *C difficile* have detectable toxin B in their faeces (toxin A was rarely estimated), often at very high titres.

This intriguing situation is not completely understood. It could be compared to protection afforded by probiotics, which we describe later. The following hypotheses can be put forward:

- **Decrease of toxin receptor**. It has been shown in rabbits (Eglow *et al.*, 1992), but not in hamsters (Rolfe, 1991) that the lower newborn sensitivity to toxin A was related to a small number of receptors. In humans the low sensitivity of enterocytes could not be proven since the receptor has not yet been identified.
- **Interactions between toxinogenic and non-toxinogenic *C. difficile***. Different authors (Wilson and Sheagren, 1983; Borriello and Barclay, 1985; Corthier and Muller, 1988) have shown that non-toxinogenic strains of *C. difficile* prevent colonization by toxinogenic strains and/ or prevent toxin production by toxinogenic *C. difficile* (see below).
- **Influence of the diet**. In gnotoxenic mice, adapted diets prevent toxin production by *C. difficile* (Mahé, Corthier and Dubos, 1986; Dubos-Ramaré and Corthier, 1990). This effect may occur in the period of milk feeding where the diet is different from adult diets.

The currently available studies show no close connection between *C.*

difficile and antibiotic-associated diarrhoea in young children, even though they are commonly colonized. However, in certain cases, the role of C. *difficile* cannot be fully ruled out. In a recent study, Buts, Corthier and Delmée (1993) described some cases of C. *difficile*-associated enteropathies in infants (19 eligible patients, median age 8 months). The patients suffered from persistent or protracted diarrhoea, malabsorption and failure to grow. C. *difficile* was the only pathogen positive for cytotoxin assay isolated in stools. No patient had evidence of colitis.

Children with chronic relapsing colitis induced by C. *difficile* toxin had significantly lower IgG anti-toxin A levels than a population of healthy children. Five (out of six) had clinical resolution of their gastrointestinal symptoms as well as clearing of C. *difficile* cytotoxin from their stools following intravenous administration of gamma-globulin (antitoxin A, Leung *et al.*, 1991). These observations suggest that a deficiency of IgG antitoxin A may predispose children to chronic relapsing C. *difficile* colitis.

3.2.5 Risk factors

As mentioned previously, any factor that can alter intestinal ecology has to be considered a risk factor. We would like to consider whether some categories of humans are more exposed than others. Patients requiring gastrointestinal surgery have to be considered as highly exposed to risk since their digestive tract is often decontaminated before surgery. It was found that pseudomembranous colitis develops more frequently in elderly persons. Is it due to the age or to the general weakness linked to other underlying pathologies? C. *difficile* was often found in patients infected by the human immunodeficiency (HIV) virus; the isolated C. *difficile* strains frequently belonged to highly pathogenic serogroups (Barbut *et al.*, 1993). These patients received large doses of antibiotics and had long hospital periods, thus intestinal disorders were frequent and they were often exposed to nosocomial C. *difficile* contamination. A similar situation was observed in patients under antineoplastic chemotherapy (review, Anand and Glatt, 1993).

3.2.6. Nosocomial epidemics

Anaerobic infections, especially those involving clostridia (sporulate organism), may be either endogenous (deriving from the host's own flora) or exogenous (resulting from exposure to pathogens from an external source). Intestinal carriage of C. *difficile* by healthy persons has been well established. Exogenous contamination is a public-health problem when it occurs in hospital (nosocomial contamination). An important study was performed on 400 patients admitted to a general medical ward over an 11-month period (McFarland *et al.*, 1989). The incidence of *C. difficile* acquisition was striking (23%). The majority of patients remained asymptomatic during the studied period. The study clearly demonstrated the importance of patient-to-patient transmission of *C. difficile* within the hospital; also substantial environmental contamination with *C. difficile* was noticed. To reduce nosocomial transmission of *C. difficile* the authors recommended appropriate hand washing, routine use of body-substance precautions for all patients and frequent environment disinfections.

3.2.7 Antibiotic treatments

Surprisingly, the treatment for pseudomembranous colitis or antibiotic-associated diarrhoea is antibiotic administration (Bartlett, 1984). Vancomycin appears to be effective against *C. difficile* disease. Oral drug administration has no side-effect and is not absorbed by the digestive tract. Nearly all patients respond to treatment. The major therapeutic alternative to vancomycin is metronidazole, which is sometimes less expensive than vancomycin. Many patients relapse after antibiotic treatment and have identical symptoms usually 3–10 days after treatment has been discontinued. Bartlett (1994) reported a relapse rate of 23%, while in other studies this percentage ranged from 5 to 55%. The presumed mechanism of relapse is based on the observation that antibiotics do not eradicate *C. difficile* and that spores survive and then revert to the vegetative form that produces toxins.

Treatment of *C. difficile* diarrhoea and/or relapses could be performed by administration with living organisms (probiotics). This point is dealt with below.

3.3 TREATMENTS BY LIVING ORGANISMS

3.3.1 Experimental animal models

Search for new treatments of disease requires the use of experimental animal models which harbour the same pathogens as humans and reproduce similar pathologies. Two models have been developed: the golden hamster treated by antibiotics and the gnotoxenic rodent. Hamsters harbour *C. difficile* in their digestive tract but bacterial growth is inhibited by the autochtonous microflora (colonization resistance; review Hentges, 1992). An antibiotic treatment with clindamycin, for instance, killed most of the microflora leading to *C. difficile* overgrowth. This model resembled the human antibiotic-associated diarrhoea. Hamsters are very sensitive to *C. difficile* toxins and die after clindamycin treatment (Bartlett, 1979; Lyerly, Krivian and Wilkins, 1988; Rolfe and Finegold, 1988). This model could be improved by inoculation of a known strain of *C. difficile* after antibiotic treatment but contaminations by other *C. difficile* strains cannot be excluded. The second animal model involved germ-free rodents. In this case, growth of *C. difficile* could not be blocked and rodents developed pseudomembranous colitis (Onderdone, Cisneros and Bartlett, 1980; Czuprinski *et al.*, 1983; Corthier, Dubos and Ducluzeau, 1986).

3.3.2 Attempts to restore colonization resistance to *C. difficile*

Antibiotic-associated diarrhoea and pseudomembranous colitis occur after destruction of colonization resistance leading to *C. difficile* overgrowth. Several authors have attempted to restore the antagonistic effect of the flora by oral administration of live bacteria.

(a) Treatment by stool in humans

Different authors have shown that colonization resistance in hamsters could be restored by ingestion of a faecal suspension. The idea led to oral treatment with suspensions of heterologous faeces in humans. Schwan, Sjolin and Trottestam (1983) described a 65-year-old patient with multiple relapses of *C. difficile* enterocolitis who received enemas (prepared in an anaerobic cabinet) with fresh faeces from her husband. This rectal infusion resulted in prompt and complete normalization of bowel function. No relapses occurred during the following 9 months. Bowden, Mansberger and Lykins (1981) tested a similar treatment (enemas of faeces obtained from surgical residents) over a 18-year period on 16 patients with pseudomembranous colitis. Thirteen of the patients responded dramatically, with decrease in diarrhoea. Of the three who died, two did not have the pseudomembranous colitis at death, and one had involvement of the small bowel (Bowden, Mansberger and Lykins, 1981). The hazard in this 'successful' procedure, of course, is the possibility of transferring other infectious or oncogenic agents (such as HIV) by means of faecal enemas.

(b) The barrier effect of the normal microflora

An alternative is to identify and cultivate the bacteria involved in colonization resistance. Several laboratories (including our own) have failed to isolate and identify such bacteria in the human flora. Some success has been obtained by a group working on the hamster caecal microflora. Three anaerobic bacterial strains consisting of a minimum microbiota, responsible for *C. difficile* colonization resistance in the gut of the hamster, were isolated (Boureau *et al.*, 1989). These strains (*Clostridium indolis, C. cocleatum* and *Eubacterium* sp.) were transferred sequentially to germ-free mice, in order to obtain monoxenic, dixenic and trixenic animals. Only trixenic mice showed *C. difficile* colonization resistance. Using scanning electron microscopy, it was shown that each step of colonization was in close association with mucus (except the mucus situated inside the crypts). The mechanism of colonization resistance might involve mucus interaction (degradation or interaction with specific receptor). The inoculation order seemed to be very important in the restoration of colonization resistance, suggesting that medical applications could be complex. Furthermore, strains involved in the human colonization resistance have to be identified. At present, this approach cannot be adapted for human treatment.

We have mentioned earlier that the colonization resistance to *C. difficile* is often disrupted by antibiotic treatment. It was shown recently in an animal model that drugs devoid of antibiotic activity could have an effect on the colonization resistance to *C. difficile* (Barc *et al.* 1994). Mice received a complex flora which prevented colonization by *C. difficile*, then an oral administration of methotrexate (agent used in long-term cancer chemotherapy) and a challenge with a toxigenic *C. difficile* strain. The *Clostridium* established in treatment animals, but was not detectable in control mice. The effects were reduced with neuroleptics (cyamemazine and chlorpromazine). These data indicate that colonization resistance to *C. difficile* could be broken by drugs needed in long-term treatments. Attempts to restore it with a mixture of faecal bacteria will probably be too difficult to guarantee a quick effect after disruption of colonization resistance. This is in favour of using probiotics – living organisms, given by the oral route, which need not establish in the gastrointestinal tract.

(c) Treatments by lactic bacteria or dairy products

For many years there have been claims that various fermented dairy products can be used to prevent or treat diarrhoea. According to Goldin and Gorbach (1992) 'the results of experiments in this area are not consistent', except probably the *Lactobacillus* strain (called GG) isolated in their laboratory. It 'colonizes the human intestinal tract and adheres more tightly to human intestinal and buccal cells than other strains of *Lactobacillus*... it elaborates an anti-microbial substance which has a broad spectrum of activity against a range of bacteria, including *C. difficile*'.

Lactobacillus GG has been administered to 32 patients suffering from recurring diarrhoea caused by antibiotic-induced *C. difficile* toxin. All 32 patients improved symptomatically after receiving *Lactobacillus* GG. Twenty-seven patients were cured by a single course of treatment based on a minimum follow-up of 2 months (Gorbach, Chang and Goldin, 1987; Goldin and Gorbach, 1992). More recently, Biller *et al.*

(1995) reported a successful treatment of four young children with multiple relapses of *C. difficile* colitis using *Lactobacillus* GG. Colombel *et al.* (1987), in a double-blind placebo-controlled study (10 healthy volunteers), reported that yoghurt with *Bifidobacterium longum* reduced erythromycin-induced gastrointestinal effects in adults. The intake of yoghurt reduced the frequency of gastrointestinal disorders in volunteers receiving erythromycin and placebo yoghurt. A sharp fall of clostridial spores was noticed in the treated group but *C. difficile* or its toxins were not characterized in this study. The effect of yoghurt (the traditional formulation without *Bifidobacterium*) was tested in the hamster model of clindamycin-induced *C. difficile* colitis (Kotz *et al.*, 1992). Mortality was 100% in yoghurt-treated animals, and all hamsters showed histological changes of severe colitis. The authors have not found any evidence to support the possible efficacy of yoghurt in the prevention of *C. difficile* colitis. In our opinion, the question of the protective role of yoghurt and/or bacteria employed in dairy products, including *Bifidobacterium*, is still open. Double-blind placebo-controlled studies have to be performed on a large number of patients before any definitive conclusion can be reached.

(d) Treatment by non-toxigenic C. difficile

There is a tendency for similar bacteria to interact competitively at the same body site. The interplay between two strains of *C. difficile*, one toxigenic and the other non-toxigenic, was studied in hamster and mouse models (Wilson and Sheagren, 1983; Borriello and Barclay, 1985; Corthier and Muller, 1988).

Cefoxitin-treated hamsters were first colonized with a non-toxigenic strain of *C. difficile* and then a toxigenic strain of *C. difficile* was administered. Toxigenic *C. difficile* was suppressed to a mean caecal population level of <0.2% of that found in control animals given only toxigenic *C. difficile* after cefoxitin treatment. The protection afforded by the non-toxigenic strain was 93%, as opposed to 21% survival of the control animals. Simultaneous administration of non-toxigenic and toxigenic *C. difficile* did not confer protection (Wilson and Sheagren, 1983). A similar experiment was performed with clindamycin-treated hamsters (Borriello and Barclay, 1985). In total, 13 of 18 protected hamsters survived for up to 27 days whereas the control animals died within 2 days. It was shown that non-toxinogenic strain of *C. difficile* adhered to gut mucosa. The authors postulated competition for ecological niches.

The strains used in the preceding studies (toxigenic and non-toxigenic) had different origins which could have influenced the antagonistic effect. Another experiment (Corthier and Muller, 1988) was performed, in a gnotoxenic mouse model, with low and/or non-toxigenic *C. difficile* strains which derived from the toxigenic one. They were inoculated prior to the toxigenic *C. difficile* strain, and prevented colonization by the toxigenic *C. difficile*. In this model the protection was 100% compared to 0% in the control animal group. Simultaneous administration of non-toxigenic and toxigenic *C. difficile* was partly protective (60%). This study confirms the previous experiments using *C. difficile* from different origins in a hamster model. The different works suggested that reduction, or absence, of toxin production gave an ecological advantage to the strain that competes with the toxigenic *C. difficile*.

A similar treatment was tested with some success in humans (Seal *et al.*, 1987). Two patients with relapsing *C. difficile* diarrhoea following metronidazole and vancomycin therapy were colonized with a non-toxigenic avirulent *C. difficile* strain given orally. Both patients appeared to respond without side-effects. This type of probiotic approach has not been developed further. It may be because no one can be sure that the non-toxigenic strains will not revert and produce toxins.

3.3.3   Modulation of toxin productions

After an antibiotic treatment there is competition between restoration of the autochthonous flora (responsible for the colonization resistance) and overgrowth of *C. difficile* (leading to disease). We have mentioned earlier that colonization resistance is not restored rapidly. The aim of a probiotic treatment is to protect against *C. difficile* disease during the period where the autochthonous microflora returns to normal values. After that, probiotics can be stopped. It has been demonstrated that some micro-organisms are able to reduce toxin production when toxigenic *C. difficile* is established at high levels in the gastrointestinal tract. These organisms have no effect on *C. difficile* growth. This point, modulation of toxin production, is discussed below.

(a)   Strains isolated from human stools

In the human colon there are five predominant bacterial genera which are all strict anaerobes. In babies, facultative anaerobes such as *Escherichia coli* and *Streptococcus* are abundant in the first week of life. Several bacterial strains have been isolated from the predominant human microflora (baby and adult) and each was inoculated to different groups of germ-free mice kept in different isolators. The animals were then challenged by a toxigenic strain of *C. difficile* (Corthier, Dubos and Raiband, 1985; Table 3.3). Some strains belonging to *E. coli*, *Enterococcus faecalis* or *Clostridium* species exerted a protective effect. They did not reduce the total number of *C. difficile* but low levels of both toxins were found. More recently an experiment was carried out with *Bacteroides* strains belonging to the *Bacteroides fragilis* group (dominant *Bacteroides* in the adult human flora) (Table 3.3). All the strains were protective and reduced toxin production. *In vitro*, no direct effect of the protective bacteria could be demonstrated on toxins. This type of protection is characterized by a modulation of toxin production in the absence of colonization resis-

Table 3.3 Mouse protection from *Clostridium difficile* challenge by prior inoculation with bacterial strains isolated from the human dominant microflora. From Corthier, Dubos and Raibaud (1985)

Strain to be tested		Mouse mortality	Dead or dying mice		Alive mice	
Name	Origin	(%)	Toxin A	Toxin B	Toxin A	Toxin B
None		100	3.2 (0.4)*	2.7 (0.2)	–	–
Escherichia coli						
E. coli 1	Baby	0	–	–	<1.4	2.1 (0.2)
E. coli 2	Baby	0	–	–	<1.4	1.9 (0.2)
E. coli 3	Baby	0	–	–	<1.4	2.7 (0.3)
E. coli 4	Baby	0	–	–	<1.4	1.6 (0.3)
Streptococcus faecalis sp.						
S. faecalis 1	Baby	50	3.1 (0.4)	2.5 (0.2)	<1.4	2.4 (0.3)
S. faecalis 2	Baby	83	3.0 (0.3)	2.9 (0.3)	ND	ND
Bifidobacterium sp.						
Bif. longum 3	Baby	17	ND	ND	<1.4	2.0 (0.2)
Bif. longum 167–4	Baby	50	3.2 (0.3)	2.8 (0.2)	<1.4	1.3 (0.5)
Bif. bifidum MP10	Baby	100	3.2 (0.4)	2.8 (0.3)	–	–
Bif. bifidum DCA	Baby	100	3.3 (0.4)	3.4 (0.3)	–	–
Bif. breve 4J	Baby	100	3.2 (0.3)	2.4 (0.4)	–	–
Bif. adolescentis D	Adult	50	3.1 (0.5)	2.3 (0.2)	<1.4	1.3 (0.5)
Bacteroides sp.						
Bacteroides sp. 1	Baby	100	2.7 (0.5)	3.2 (0.3)	–	–
Bacteroides sp. 2	Baby	100	3.0 (0.4)	3.4 (0.2)	–	–
Bact. fragilis 1	Adult	0	–	–	<1.4	2.3 (0.4)
Bact. distasonis PR37	Adult	10	ND	ND	<1.4	2.2 (0.5)
Bact. vulgarus 1	Adult	0	–	–	<1.4	2.2 (0.4)
Clostridium sp.						
Clostridium sp. D1	Adult	0	–	–	<1.4	0.8 (0.7)
Clostridium sp. D10	Adult	67	3.0 (0.3)	2.6 (0.2)	<1.4	2.5 (0.2)
Clostridium sp. A1	Adult	83	3.2 (0.4)	2.6 (0.3)	ND	ND
Clostridium sp. A7	Adult	100	2.9 (0.4)	2.6 (0.3)	–	–

ND, not determined. * SEM.

tance to *C. difficile*. A similar phenomenon was observed when gnotoxenic mice were fed an appropriate diet before *C. difficile* challenge (Mahé, Corthier and Dubos, 1987).

(b) Bifidobacterium

Several *Bifidobacterium* strains were isolated from human faecal samples. The strains were inoculated 4 days before challenge by the toxigenic strains of *C. difficile* (Table 3.3, Corthier, Dubos and Raibaud, 1985). The protection was strain dependent. As in the previous case, the protective *Bifidobacterium* strain reduced the quantities of toxins detectable in the digestive tract without affecting growth of *C. difficile*. All the *Bifidobacterium bifidum* strains were protective but the number of strains tested was too low to ascertain that this *Bifidobacterium* species will always be protective. In France, new fermented milks containing *Bifidobacterium* have been commercially available since 1985. The incorported *Bifidobacterium* is supposed to have a beneficial effect on health. It will be of interest to test this hypothesis in the different antibiotic-associated diarrhoea models.

(c) Search for a mechanism of action

The mechanism of toxin modulation by probiotics is still unknown but different approaches could be developed. Fuller (1992) proposed four mechanisms of probiotic effects:

- **Competition for nutrient**. *In vitro*, regulation of *C. difficile* toxin production was observed according to amino-acid composition (Haslam *et al.*, 1986; Osgood, Wood and Sperry, 1993) and sugar origin (Kazamias and Sperry, 1995). The protective bacteria may consume 'key' products, such as arginine (Osgood, Wood and Sperry, 1993), essential for cytotoxin production. The observation that adapted diet reduces toxin production (Mahé, Corthier and Dubos, 1986; Dubos-Ramaré and Corthier, 1990) strengthens the requirement of a nutrient essential for toxin production.
- **Competition for adhesion receptors on the gut epithelium**. This hypothesis has not been tested yet.
- **Production of antitoxin substances**. *C. difficile* toxins are sensitive to proteases. However, in *in vitro* studies, toxin quantities are not reduced when incubated with protective bacteria. An indirect effect cannot be excluded.
- **Stimulation of immunity**. This hypothesis is not consistent with the observed effects, protection occurring too rapidly (minimum time: 7 days).

3.3.4   Effect of a living yeast: *Sac. boulardii*

Saccharomyces boulardii is a mesophilic, non-pathogenic yeast used in many countries to prevent diarrhoea and other gastrointestinal diseases associated with the use of antibiotics (review McFarland and Bernasconi, 1993). It was first isolated from lychees in Indochina and in France has been used to treat diarrhoea from the 1950s. It has an unusually high optimal growth temperature of 37°C. It was used successfully in animal models to treat antibiotic-associated diarrhoea and pseudomembranous colitis. These encouraging results have led to human trials which have shown that *Sac. boulardii* could be efficiently used to treat relapse of *C. difficile* and some cases of antibiotic-associated diarrhoea. The yeast is well suited as a treatment agent as it is able to achieve high concentrations in the colon quickly, maintain constant levels, does not permanently colonize the colon and does not translocate easily out of the intestinal tract (McFarland and Bernasconi, 1993). Moreover, being a yeast, *Sac. boulardii* is genetically resistant to antibacterial antibiotics. *Sac. boulardii* must be considered as a pharmaceutical probiotic.

(a)   The animal model

Saccharomyces boulardii has been tested in animal models of *C. difficile*-associated colitis and has been found to have a protective effect against toxigenic *C. difficile* challenge. Efficacy of *Sac. boulardii* was first demonstrated in clindamycin-treated hamsters (Massot *et al.*, 1984; Toothaker and Elmer, 1984; Elmer and McFarland, 1987). If hamsters were treated with *Sac. boulardii* from the final day of antibiotic treatment, they were largely protected from mortality. While hamsters protected with the yeast still harboured relatively high viable counts of *C. difficile* in their caeca, a sharp reduction in toxin positivity was observed.

Corthier, Dubos and Ducluzeau (1986) found that gnotoxenic mice, who usually died after a *C. difficile* challenge, were protected after a single dose of *Sac. boulardii* (16% survival) and 56% survived if the *Sac. boulardii* was given continuously in drinking water. As reported in a previous experiment, no direct antagonistic effect of the yeast on *C. difficile* numbers was detected whereas a modulation of faecal toxin production was observed. In caecal contents of dead or moribund animals about $1\,\mu g$ (g^{-1} of caecal content) of toxins A and B was detected whereas the caecal quantities in protected mice were lower than $1\,ng\,g^{-1}$ for both toxins although *C. difficile* was still present at high levels ($> 10^8\,c.f.u.\,g^{-1}$) (Corthier, Dubos and Raibaud, 1986; Elmer and Corthier, 1991). Using scanning electron microscopy, it was observed that the digestive tract of mice protected by yeast was normal, compared to the control group where the epithelium was completely destroyed (Castex *et al.*, 1990). The protection from *C. difficile*-associated colitis was later found to be dependent upon the dose and viability of the yeast. The ability of *Sac. boulardii* to inhibit *C. difficile*-associated damage is lost if yeast is given in a non-viable state (killed by heat or amphotericin B). In addition, a dose response was observed in a study by Elmer and Corthier (1991). As the dose of *Sac. boulardii* was increased from 3×10^8 to 3.3×10^{10} c.f.u. ml^{-1} in drinking water given to mice, survival increased linearly from 9 to 85%. After 1 week, the yeast could be removed from the drinking water of protected mice without any further development of disease (Corthier, Dubos and Ducluzeau, 1986). The toxigenic *C. difficile* strain dissociated, in the gastrointestinal tract, into non-toxigenic or low-toxigenic clones. These clones reduced toxin production by the strongly toxigenic *C. difficile* (Corthier and Muller, 1988; see above).

(b) Clinical use of Sac. boulardii

Prevention of antibiotic-associated diarrhoea

Three large studies have been completed using *Sac. boulardii* as a preventive therapy for antibiotic-associated diarrhoea. The first focused on ambulatory patients in a multicentric study in France involving 25 centres with a total of 388 enrolled patients (Adam, Barret and Barret-Bellet, 1977). Patients aged 15 years or older who received tetracycline or beta-lactam antibiotics for at least 5 days and had no intestinal pathology were enrolled in a double-blind, placebo-controlled trial. Of the 199 patients treated with *Sac. boulardii*, significantly fewer (4.5%) developed antibiotic-associated diarrhoea compared to patients receiving placebo (17.5%). The second double-blind study (Surawicz *et al.*, 1989a, b) was carried out on patients receiving new antibiotic prescriptions in a hospital in Seattle, WA. *Sac. boulardii* or a placebo was assigned as a concurrent therapy for the antibiotics. The study drug was started within 2 days of antibiotic initiation and continued for 2 weeks after antibiotics were discontinued. *Sac. boulardii* was given at a dose of 500 mg b.i.d. for a total of 1 g day^{-1}. Among the 180 tested patients, 14 of the 64 (21.8%) given the placebo developed antibiotic-associated diarrhoea compared with 11 of the 116 (9.5%) patients on *Sac. boulardii*. In this study, there were no significant adverse reactions related to *Sac. boulardii* exposure. The third study (McFarland *et al.*, 1994) tested *Sac. boulardii* as a preventive agent for hospitalized patients receiving at least one beta-lactam antibiotic (alone or in conjunction with other antibiotics). Patients were blindly assigned to either *Sac. boulardii* (500 mg b.i.d.) or a placebo at least 75 h after beta-lactam was begun and continued for at least 2 days after the antibiotic was discontinued. Among the 181 eligible patients, 6/89 (6.7%) on *Sac. boulardii* developed antibiotic-associated diarrhoea, which was significantly fewer than the 14/92 (15.2%) of the patients on the placebo who developed antibiotic-associated diarrhoea. As in the previous study, among the 23 patients who were positive for *C. difficile*, the frequency of diarrhoea was lower in patients on *Sac. boulardii* (20%) compared with patients on the placebo (31%). The results of these trials indicate that *Sac. boulardii* is an effective pharmaceutical probiotic for the prevention of antibiotic-associated diarrhoea.

Treatment of relapsing *C. difficile*-associated diarrhoea
Thirteen patients who had history of recurrent disease were treated with vancomycin (500 mg day^{-1}) for 10 days and a 30-day course of *Sac. boulardii* (1 g day^{-1}). Eleven (85%) responded successfully to the combination of vancomycin and *Sac. boulardii* and did not experience any further recurrence of disease (Surawicz *et al.*, 1989a). The next study was a double-blind placebo-controlled trial of 102 patients who had either recurrent episodes of *C. difficile* or were experiencing *C. difficile* disease for the first time. Patients were given *Sac. boulardii* or placebo with at least 4 overlapping days of vancomycin or metronidazole treatment. The study drug was continued for 28 days (1 g day^{-1}). The patients were studied for an additional 28 days after drug discontinuation to observe for recurrence of *C. difficile* disease. Of the 51 patients with incident (first-time) *C. difficile* disease, seven of 29 (24%) patients on the placebo had a recurrence and two of 22 (9%) patients on *Sac. boulardii*. Of the 51 patients with a history of recurrent *C. difficile* disease, 19 of 28 (68%) patients on the placebo had another recurrence but only nine of 23 (39%) of patients on *Sac. boulardii* reported a recurrence. Overall, 11 of 45 (24%) patients on *Sac. boulardii* experienced a clinical recurrence and significantly more patients on the placebo had a further recurrence (26 of 57 – 45%). Another study on children was performed by Buts, Corthier and Delmée (1993). Children aged 3 months to 11 years presented enteral symptoms lasting for more than 15 days and had *C. difficile* toxin-B-positive stools. The treatment with *Sac. boulardii* (250 mg, two to four times day^{-1}) resulted in rapid improvement of enteral symptoms (95%) and in clearing of stool toxin B by day 15 in 16 cases (85%). A clinical relapse occurred in two patients which resolved rapidly with a second *Sac. boulardii* treatment.

(c) Mechanisms of actions

All the studies concluded that *Sac. boulardii* does not exert an antagonistic effect against *C. difficile* overgrowth. The remaining question is how does *Sac. boulardii* protect against *C. difficile* disease: what is (are) the mechanism(s) of this pharmaceutical probiotic? There is probably no 'unique' effect but different mechanisms which help and/or compete with each other. The four hypotheses of probiotic effects proposed by Fuller (1992) were analysed as described below.

Competition for nutrients
We have discussed previously that such competition was the more likely explanation of the modulation of toxin production observed with protective diets or bacteria. In a gnotobiotic mouse model, it was shown that yeast treatment reduces the quantities of free toxins in the gastrointestinal tract (Corthier, Dubos and Ducluzeau, 1986; Elmer and Corthier, 1991) but no direct effect of *Sac. boulardii* on toxin activities could be demonstrated (Corthier *et al.*, 1992), suggesting that the yeast changes a signal in toxin regulation leading to a decrease in toxin production. This 'key' signal received by *C. difficile* has not yet been identified. Further work on toxin regulation in *C. difficile* must be performed before this can be achieved.

Competition for adhesion receptors on the gut epithelium
In an attempt to understand the mechanism of the protective effect, the action of *Sac. boulardii* on crude toxin preparation was studied under *in vitro* and *in vivo* conditions. Incubation of toxins with *Sac. boulardii* prior to oral inoculation into conventional mice did not protect them. An oral *Sac. boulardii* treatment (duration 4 days) given before oral toxin inoculation was protective. The mucosa of *Sac. boulardii*-protected mice were not damaged (scanning electron microscopy), suggesting that the yeast may act on the intestinal mucosa (Corthier *et al.*, 1992).

Production of antitoxin and/or antireceptor substances
Modification of the brush-border receptor is probably one of the mechanisms of *Sac. boulardii* protection. Czerucka *et al.* (1991) found a decrease in the percentage rounding of intestinal cells due to *C. difficile* if *Sac. boulardii* was added to cell culture. Other experiments were performed with purified toxin A (^{3}H radiolabelled) inoculated to ligated rat ileal loop (Pothoulakis *et al.*, 1993). *Sac. boulardii* reduced toxin–receptor binding in a dose-dependent fashion. SDS-PAGE of ileal brush border exposed to *Sac. boulardii* revealed a diminution of all brush-border proteins. Treatment of rats with *Sac. boulardii* suspension reduced fluid secretion and mannitol permeability caused by toxin A. The authors postulated the existence of a 54 kd serin-protease secreted by *Sac. boulardii*. The same research group identified a trypsin-like protease from *Sac. boulardii* culture. This protease inhibited toxin A receptor binding and enterotoxicity. Pretreatment of toxin A with purified protease partially digested the toxin A *in vitro* and reduced fluid secretion, permeability and morphological damage in ligated ileal loops (Castagliuolo *et al.*, 1993). Buts, De Keyser and De Raedemaeker (1994) also showed that *Sac. boulardii* enhanced enzyme expression by rat enterocytes. This effect could lead to production of antitoxin substances. Oral administration of *Sac. boulardii* (1 g day^{-1} for 14 days) was shown by Buts *et al.* (1986) to increase the duodeno-jejunum levels of mucosal sucrase (82% increase),

lactase (77%) and maltase (75%) in seven human volunteers. An increase in these three enzymes was also found in rats fed *Sac. boulardii* (75 mg day^{-1} for 5 days) compared with controls. This increased activity may modify toxin A receptor.

Stimulation of immunity
In antibiotic-associated diarrhoea, yeast protection occurs too rapidly to be due to an immune process but it has been proved that *Sac. boulardii* stimulated immunity. Oral ingestion of *Sac. boulardii* caused an increase in secretary IgA and secretory component in the small intestine of rats. Buts *et al.* (1990) found that sucking and weaning rats given high doses of *Sac. boulardii* (0.5 mg g^{-1} t.i.d.) had a 80% increase in secretory component in the crypt cells over controls and 69% increase in secretory component in villus cells. The mean secretory IgA level was increased by 57% in *Sac. boulardii*-treated rats compared with controls given saline.

3.4 SEARCH FOR NEW ORAL TREATMENTS BY PROBIOTICS

At present, only the yeast *Sac. boulardii* has been clearly shown to be successful in the treatment of antibiotic-associated diarrhoea in humans. Experimental and clinical data on other micro-organisms such as *Lactobacillus* or *Bifidobacterium* are interesting but their therapeutic use needs further investigation. In the future, the use of genetically modified organisms must be considered. To conclude, we now discuss what kind of new probiotics could be imagined to treat antibiotic-associated diarrhoea.

3.4.1 Oral vaccination

Immunity was found to be efficient to protect from *Clostridium difficile* enteropathies. Passive immunity could protect against *C. difficile* disease (Fernie *et al.*, 1983; Lyerly *et al.*, 1991). Monoclonal antibodies were prepared using toxin A as antigen (Corthier *et al.*, 1991). The antibodies were specific of the repeat C terminal portion. Mice were fully protected by a systemic transfer of the immunoglobulin. Vaccines have been tested, in animal models, with mixtures of inactivated toxins (Lyerly, Krivian and Wilkins, 1988) or synthetic vaccines. Lyerly *et al.* (1990) developed a vaccine with a non-toxic recombinant peptide comprising 33 of the 38 repeating units of toxin A. The antisera neutralized the enterotoxic and cytotoxic activity of the toxin and the hamsters were partially protected against *C. difficile* disease. Warny *et al.* (1994) showed that the clinical course of *C. difficile* infection may be related to a weak immune response. Five children were treated with intravenously administered gamma globulin (Leung *et al.*, 1991). All of them had clinical resolution of their gastrointestinal symptoms as well as clearing of *C. difficile* cytotoxin B from their stools. A successful treatment with oral IgA supplement was observed in a case of severe diarrhoea resistant to other therapies (Tjellstrom *et al.*, 1993). All these studies indicated that an immune response to toxin could protect from disease. Frey and Wilkins (1992) identified a small epitope on toxin A. The door is open for synthetic vaccines with a non-toxic and immunogenic part of the molecule. The future for probiotics could be to produce synthetic antigens in the digestive tract to enhance the immune response.

3.4.2 Modulation of toxin production

It has been mentioned previously that different factors (adapted diet, bacteria isolated from human predominant microflora and *Sac. boulardii*) reduced toxin production by *C. difficile* in the gastrointestinal tract. To understand these observations, fundamental research has to be performed on the regulation of toxin production. This type of work is in progress in different countries (including the USA, England, Germany and France). The coding sequences for both toxins are known and different promoter sequences have been suggested (Barroso *et al.*, 1984; Dove, Wang and Price, 1990; Eichel-Streiber *et al.*, 1992). The techniques for gene transfer in *C. difficile* has been improved recently (Mullany *et al.*, 1990). All these studies suggest that tools exist to determine how *C. difficile* toxin production is regulated (more studies may have been carried out but are yet to be published). The 'key' factors could be activators or repressors of gene transcriptions. When they are characterized, the field will be open for a new kind of probiotics. These living organisms will have to modify the gut content in order to prevent toxin production (adding or removing the 'key' molecule). To be safe, this treatment will have to be employed preventively so as to reduce the risks in populations exposed to *C. difficile* disease, such as elderly persons or patients treated for acquired immunodeficiency syndrome (AIDS) or cancer.

3.5 ACKNOWLEDGEMENT

The authors wish to thank A. Bouroche for her help in improving the English used in this chapter.

3.6 REFERENCES

Adam, J., Barret, A. and Barret-Bellet, C. (1977) Essais cliniques contrôlés en double insu de l'ultralevure lyophilisée. Etude multicentrique par 25 médecin de 388 cas. *Gazette Med. France*, **84**, 2072–8.

Anand, A. and Glatt, A.E. (1993) *Clostridium difficile* infection associated with antineoplastic chemotherapy: a review. *Clin. Infect. Dis.*, **17**, 109–13.

Barbut, F., Depitre, C., Delmée, M. *et al.* (1993) Comparison of enterotoxin production, cytotoxin production, serogrouping and antimicrobial susceptibilities of *Clostridium difficile* strains isolated from AIDS and Human Immunodeficiency Virus-Negative patients. *J. Clin. Microbiol.*, **31**, 740–2.

Barc, M.C., Depitre, C., Corthier, G. *et al.* (1994) Barrier effect of normal microbiota against *Clostridium difficile* may be influenced by drugs devoid of antibiotic activity. *Microb. Ecol. Health Dis.*, **7**, 307–13.

Barroso, L.A, Wang, S.Z, Phelps, C.J, *et al.* (1984) Nucleotide sequence of *Clostridium difficile* toxin B gene. *Nucleic Acids Res.* **18**, 4004.

Bartlett, J.G. (1979) Antibiotic-associated-pseudomembranous colitis. *Rev. Infect. Dis.*, **1**, 530–8.

Bartlett, J.G. (1984) Treatment of antibiotic associated pseudomembranous colitis. *Rev. Infect. Dis.*, **6**, (Suppl. 1), S235–41.

Barlett, J.G. (1994) *Clostridium difficile*: history of its role as an enteric pathogen and the current state of knowledge about the organism. *Clin. Infect. Dis.*, **6**, (Suppl. 4), S265–72.

Benoit, R., Danquechin-Dorval, E., Loulergue, J. *et al.* (1992) Diarrhée post-antibiotique: rôle de *Klebsiella oxytoca*. *Gastroenterol. Clin. Biol.*, **16**, 860–4.

Biller, J.A., Katz, A.J., Flores, A.F. *et al.* (1995) Treatment of recurrent *Clostridium difficile* colitis with *Lactobacillus* GG. *J. Pediatr. Gastroenterol. Nutr.*, **21**, 224–6.

Borriello, S.P. and Barclay, F.E. (1985) Protection of hamsters against *Clostridium difficile* ileocaecitis by prior colonisation with non pathogenic strains. *J. Med. Microbiol.*, **19**, 339–50.

Boureau, H., Decre, D., Popoff, M. *et al.* (1989) Isolation and identification of microflora resistant to colonization by *Clostridium difficile*. *Microecol. Ther.*, **18**, 117–20.

Bowden, J.T.A, Mansberger, J.A.R. and Lykins, L.E. (1981) Pseudomembranous enterocolitis: mechanism of restoring floral homeostasis. *Am. Surgeon*, **47**, 178–83.

Buts, J.P, Corthier, G. and Delmée, M. (1993) *Saccharomyces boulardii* for *Clostridium difficile* associated enteropathies in infants. *J. Pediatr. Gastroenterol. Nutr.*, **16**, 419–25.

Buts, J.P, De Keyser, N. and De Raedemaeker, L. (1994) *Saccharomyces boulardii* enhances rat intestinal enzyme expression by endoluminal release of polyamines. *Pediatr. Res.*, **36**, 522–7.

Buts, J.P, Bernasconi, P., Craynest, M. *et al.* (1986) Response of human and rat intestinal mucosa to oral administration of *Saccharomyces boulardii*. *Pediatr. Res.*, **20**, 192–6.

Buts, J.P., Bernasconi, P., Vaerman, J.P. and Dive, C. (1990) Stimulation of secretory IgA and secretory component of immunoglobulins in small intestine of rats treated with *Saccharomyces boulardii*. *Dig. Dis. Sci.*, **35**, 251–6.

Castagliuolo, I., Papadimitriu, G., Jaffer, A. *et al.* (1993) *Saccharomyces boulardii* secretes a protease which inhibits *Clostridium difficile* toxin A receptor binding and enterotoxicity. *Gastroenterology.*, **104**, (Suppl. 4), A678.

Castagliuolo, I., LaMont, J.T., Letourneau, R. *et al.* (1994) Neuronal involvement in the intestinal effects of *Clostridium difficile* toxin A and *Vibrio cholerae* enterotoxin in rat ileum. *Gastroenterology*, **107**, 657–65.

Castex, F., Corthier, G., Jouvert, S. *et al.* (1990) Prevention of *Clostridium difficile* induced experimental pseudomembranous colitis by *Saccharomyces boulardii*: a scanning electron microscopic and microbiological study. *J. Gen. Microbiol.*, **136**, 1085–9.

Colombel, J.F., Cortot, A., Neut C. and Romond, C. (1987) Yogourt with *Bifidobacterium longum* reduces erythromycin-induced gastrointestinal effects. *Lancet*, **ii**, 43.

Corthier, G. and Muller, M.C. (1988) Emergence in gnotobiotic mice of non toxinogenic clones of *Clostridium difficile* from a toxinogenic one. *Infect. Immun.*, **56**, 1500–4.

Corthier, G., Dubos, F. and Raibaud, P. (1985) Modulation of cytotoxin production by *Clostridium difficile* in the intestinal tract of gnotobiotic mice inoculated with various human intestinal bacteria. *Appl. Envir. Microbiol.*, **49**, 250–2.

Corthier, G., Dubos, F. and Raibaud, P. (1986) Ability of 2 *Clostridium difficile* strains from man and hare to produce cytotoxin in vitro and in gnotobiotic rodent intestine. *Ann. Inst. Pasteur/Microbiol.*, **137B**, 113–21.

Corthier, G., Dubos, F. and Ducluzeau, R. (1986) Prevention of *Clostridium difficile* induced mortality in gnotobiotic mice by *Saccharomyces boulardii*. *Can. J. Microbiol.*, **32**, 894–6.

Corthier, G., Muller, M.C., Elmer, G.W. *et al.* (1989) Interrelationships between digestive proteolytic activities and production and quantitation of toxins in pseudomembranous colitis induced by *Clostridium difficile* in gnotobiotic mice. *Infect. Immun.*, **57**, 3922–7.

Corthier, G., Muller, M.C., Wilkins, T.D. *et al.* (1991) Protection against experimental pseudomembranous colitis in gnotobiotic mice by use of monoclonal antibodies against *Clostridium difficile* toxin A. *Infect. Immun.*, **59**, 1192–5.

Corthier, G., Lucas, F., Jouvert, S. and Castex, F. (1992) Effect of oral *Saccharomyces boulardii* treatment on the activity of *Clostridium difficile* toxins in mouse digestive tract. *Toxicon*, **30**, 1583–9.

Czerucka, D., Nano, J.L, Bernasconi, P. and Rampal, P. (1991) Response of the IRD intestinal epithelial cell line to *Clostridium difficile* toxins A and B in rats. Effect of *Saccharomyces boulardii*. *Gastroenterol. Clin. Biol.*, **15**, 22–7.

Czuprinski, C.J., Johnson, W.J., Balish, E. and Wilkins, T.D. (1983) Pseudomembranous colitis in *Clostridium difficile* monoassociated rats. *Infect. Immun.*, **39**, 1368–76.

Dove, C.H., Wang, S.Z. and Price, S.B. (1990) Molecular characterization of the *Clostridium difficile* toxin A gene. *Infect. Immun.*, **58**, 480–8.

Dubos-Ramaré, F. and Corthier, G. (1990) Influence of dietary proteins on production of *Clostridium difficile* toxins in gnotobiotic mice. *Microb. Ecol. Health Dis.*, **3**, 231–4.

Eglow, R., Pothoulakis, C., Itzkowitz, S. *et al.* (1992) Diminished *Clostridium difficile* toxin A sensitivity in newborn rabbit ileum is associated with decreased toxin A receptor. *J. Clin. Invest.*, **90**, 822–9.

Eichel-Streiber Von, C., Laufenberg, C., Feldmann, R. *et al.* (1992) Comparative sequence analysis of the *Clostridium difficile* toxins A and B. *Mol. Gen. Genet.*, **233**, 260–8.

Elmer, G.W. and McFarland, L.V. (1987) *Saccharomyces boulardii* suppression of overgrowth of toxigenic *Clostridium difficile* following Vancomycin treatment in the hamster. *Antimicrob. Agents Chemother.*, **31**, 129–31.

Elmer, G.W and Corthier, G. (1991) Modulation of *Clostridium difficile* induced mortality as a function of the dose and the viability of the *Saccharomyces boulardii* used as a preventative agent in gnotobiotic mice. *Can. J. Microbiol.*, **37**, 315–17.

Frey, S.M. and Wilkins, T.D. (1992) Localization of two epitopes recognized by monoclonal antibody PCG4 on *Clostridium difficile* toxin A. *Infect. Immun.*, **60**, 2488–92.

Fuller, R. (1992) Problems and prospects, in *Probiotics: the Scientific Basis* (ed. R. Fuller), Chapman & Hall, London, 377–86.

Goldin, B.R. and Gorbach, S.L. (1992) Probiotics for humans, in *Probiotics: the Scientific Basis* (ed. R. Fuller), Chapman & Hall, London, 355–76.

Gorbach, S.L., Chang, T.W. and Goldin, B.R. (1987) Successful treatment of relapsing *Clostridium difficile* colitis with *Lactobacillus* GG. *Lancet*, **ii**, 1519.

Hall, I.C. and O'Toole, E. (1935) Intestinal flora in new-born infants. *Am. J. Dis. Child.*, **49**, 390–402.

Haslam, S.C., Ketley, J.M., Mitchell, T.J. *et al.* (1986) Growth of *Clostridium difficile* and production of toxins A and B in complex and defined media. *J. Med. Microbiol.*, **21**, 293–7.

Hecht, G., Pothoulakis, C., LaMont, J.T. and Madara, J.L. (1988) *Clostridium difficile* toxin A perturbs cytoskeletal structure and tight junction permeability of cultured human intestinal epithelial monolayers. *J. Clin. Invest.*, **82**, 1516–24.

Hentges, D.J. (1992) Gut flora and disease resistance, in *Probiotics: the Scientific Basis* (ed. R. Fuller), Chapman & Hall, London, 87–110.

Just, I., Selzer, J., Eichel-Streiber Von, C. and Aktories, K. (1995) The low molecular mass GTP-binding protein Rho is affected by toxin A from *Clostridium difficile*. *J. Clin. Invest.*, **95**, 1026–31.

Karjalainen, T., Barc, M.C., Collignon, A. *et al.* (1994) Cloning of a genetic determinant from *Clostridium difficile* involved in adherence to tissue culture cells and mucus. *Infect. Immun.*, **62**, 4347–55.

Kazamias, M.T. and Sperry, J.F. (1995) Enhanced fermentation of mannitol and release of cytotoxin by *Clostridium difficile* in alkaline culture media. *Appl. Environ. Microbiol.*, **61**, 2425–7.

Kotz, C.M, Peterson, L.R, Moody, J.A. *et al.* (1992) Effect of yogurt on clindamycin induced *Clostridium difficile* colitis in hamsters. *Dig. Dis. Sci.*, **37**, 129–32.

Leung, D.Y.M., Kelly, C.P., Boguniewicz, M. *et al.* (1991) Treatment with intraveously administered gamma globulin of chronic relapsing colitis induced by *Clostridium difficile* toxin. *J. Pediatr*, **118**, 633–7.

Lyerly, D.M., Krivian, H.C. and Wilkins, T.D. (1988) *Clostridium difficile*: its disease and toxins. *Clin. Microbiol. Rev.*, **2**, 1–118.

Lyerly, D.M., Johnson, J.L., Frey, S.M. and Wilkins, T.D. (1990) Vaccination against lethal enterocolitis with a nontoxic recombinant peptide of toxin A. *Curr. Microbiol.*, **21**, 29–32.

Lyerly, D.M., Bostwick, E.F., Binion, S.B. and Wilkins, T.D. (1991) Passive immunisation of hamsters against disease caused by *Clostridium difficile* by use of bovine immunoglobulin G concentrate. *Infect. Immun.*, **59**, 2215–18.

Mahé, S., Corthier, G. and Dubos, F. (1986) Effect of various diets on toxin production by two strains of *Clostridium difficile* in gnotobiotic mice. *Infect. Immun.*, **55**, 1801–5.

Martinez, R.D. and Wilkins, T.D. (1992) Comparison of *Clostridium sordellii* toxins HT and LT with toxins A and B of *Clostridium difficile*. *J. Med. Microbiol.*, **36**, 30–6.

Massot, J., Sanchez, O., Couchy, R. *et al.* (1984) Bacterio-pharmacological activity of *Saccharomyces boulardii* in clindamycin induced colitis in the hamster. *Arzneim Forsch.*, **34**, 794–7.

McFarland, L.V. and Bernasconi, P. (1993) *Saccharomyces boulardii*: a review of an innovative biotherapeutic agent. *Microb. Ecol. Health Dis.*, **6**, 157–71.

McFarland, L.V., Mulligan, M.E., Kwok, R.Y. and Stamm, W.E. (1989) Nosocomial acquisition of *C. difficile* infection. *N. Engl. J. Med.*, **320**, 204–10.

McFarland, L.V., Surawicz, C.M., Greenberg, R.N. *et al.* (1994) A randomized placebo-controlled trial of *Saccharomyces boulardii* in combination with standard antibiotics for *Clostridium difficile* disease. *J. Am. Med. Assoc.*, **271**, 1913–18.

Minami, J., Saita, S., Yoshida, T. *et al.* (1992) Biological activities and chemical composition of a cytotoxin of *Klebsiella oxytoca*. *J. Gen. Microbiol.*, **138**, 1921–7.

Mullany, P., Wilks, M., Lamb, I. *et al.* (1990) Genetic analysis of a tetracycline resistance element from *Clostridium difficile* and its conjugal transfer to and from *Bacillus subtilis*. *J. Gen. Microbiol.*, **136**, 1343–9.

Onderdonk, A.B., Cisneros, R.L.J. and Bartlett, G. (1980) *Clostridium difficile* in gnotobiotic mice. *Infect. Immun.*, **28**, 277–82.

Osgood, D.P., Wood, N.P. and Sperry, J.F. (1993) Nutritional aspects of cytotoxin production by *Clostridium difficile*. *Appl. Environ. Microbiol.*, **59**, 3985–8.

Popoff, M.R. (1987) Purification and characterization of *Clostridium sordellii* lethal toxin and cross-reactivity with *Clostridium difficile* cytotoxin. *Infect. Immun.*, **55**, 35–43.

Pothoulakis, C., LaMont, J.T., Eglow, R. *et al.* (1991) Characterization of rabbit ileal receptors for *Clostridium difficile* toxin A. Evidence for a receptor coupled G protein. *J. Clin. Invest.*, **88**, 119–25.

Pothoulakis, C., Kelly, C.P., Joshi, M.A. *et al.* (1993) *Saccharomyces boulardii* inhibits *Clostridium difficile* toxin A binding and enterotoxicity in rat ileum. *Gastroenterology*, **104**, 1108–15.

Riegler, M., Sedivy, R., Pothoulakis, C. *et al.* (1995) *Clostridium difficile* toxin B is more potent than toxin A in damaging human colonic epithelium *in vitro*. *J. Clin. Invest.*, **95**, 2004–11.

Rolfe, R.D. (1991) Binding kinetics of *Clostridium difficile* toxins A and B to intestinal brush border membranes from infant and adult hamsters. *Infect. Immun.*, **59**, 1223–30.

Rolfe, D. and Finegold, S.M. (eds) (1988) *Clostriduim Difficile: Its Role in Intestinal Disease*, Academic Press, New York, 1–390.

Schwan, A., Sjolin, S. and Trottestam A. (1983) Relapsing *Clostridium difficile* enterocolitis cured by rectal infusion of homologous faeces. *Lancet*, **2**, 845.

Seal, D., Borriello, S.P., Barclay, F. *et al.* (1987) Treatment of relapsing *Clostridium difficile* diarrhea by administration of a non-toxinogenic strain. *Eur. J. Clin. Microbiol.*, **6**, 223–9.

Surawicz, C.M., McFarland, L.V., Elmer, G.W. and Chinn, J. (1989a) Treatment of recurrent *Clostridium difficile* colitis with vancomycin and *Saccharomyces boulardii*. *Am. J. Gastroenterol.*, **84** 1285–7.

Surawicz, C.M., Elmer, G.W., Speelman, P. *et al.* (1989b) Prevention of antibiotic-associated diarrhea by *Saccharomyces boulardii*: a prospective study. *Gastroenterology*, **96**, 981–8.

Tjellstrom, B:, Stenhammar, L., Eriksson, S, and Magnusson, K.E. (1993) Oral immunoglobulin A supplement in treatment of *Clostridium difficile* enteritis. *Lancet*, **341**, 701–2.

Toothaker, R.D. and Elmer, G.W. (1984) Prevention of clindamycin induced mortality in hamster by *Saccharomyces boulardii*. *Antimicrob. Ag. Chemother.*, **26**, 552–6.

Warny, M., Vaerman, J.P, Avesani, V. and Delmée, M. (1994) Human antibody response to *Clostridium difficile* toxin A in relation to clinical course of infection. *Infect. Immun.*, **62**, 384–9.

Wilson, K.H. and Sheagren J.N. (1983) Antagonism of toxigenic *Clostridium difficile* by nontoxigenic *Clostridium difficile*. *J. Infect. Dis.*, **147**, 733–6.

CHAPTER 4

Lactose maldigestion

P. Marteau, T. Vesa and J. C. Rambaud

4.1 INTRODUCTION

Lactose, the main milk sugar, is frequently partly maldigested either because of gastrointestinal diseases or because of the physiological decline of the intestinal lactase activity with age. Lactose maldigestion can lead to clinical symptoms of 'lactose intolerance'. Improvement of lactose digestion and tolerance is one of the best established properties of probiotics, especially lactic acid bacteria. In this chapter we review the basic mechanisms of lactose digestion and tolerance, evidence for the good digestion and tolerance of fermented dairy products by lactase deficient subjects, and the mechanisms involved in this effect.

4.2 LACTOSE METABOLISM

4.2.1 Lactose digestion

(a) Description of lactose digestion

Lactose, the milk sugar, is a disaccharide found only in milk. It is composed of galactose and glucose joined by a 1,4-linkage. The lactose content of human milk is on average $70\,\mathrm{g\,l^{-1}}$ and that of cow's milk $45\,\mathrm{g\,l^{-1}}$ (for lactose contents of dairy products see Scrimshaw and Murray, 1988). To be digested, lactose must first be hydrolysed by a lactase into glucose and galactose, which are actively absorbed. The main intestinal lactase, lactase-phlorizin hydrolase (LPH) (EC 3.2.1.23-3.2.1.62), is an integral glycoprotein of the brush border which is expressed specifically in the enterocytes of the upper small intestine (Semenza and Auricchio, 1995). After synthesis, LPH is transformed in

Probiotics 2: Applications and practical aspects.
Edited by R. Fuller.
Published in 1997 by Chapman & Hall. ISBN 0 412 73610 1

the cell by glycosylation and partial proteolysis (post-translational mechanisms) and then transferred to the outer surface of the brush border. High LPH activity is found in the jejunum, whereas the duodenum and ileum show low or no activity. LPH is expressed on the membrane of the enterocytes located along the villus but not on the younger crypt cells. While LPH is expressed by all enterocytes of the jejunum villi in suckling animals, and 'lactase persistent' adults, it has a patchy pattern of expression (i.e. only a few enterocytes on a villus express lactase) in the jejunum of adults with primary hypolactasia. LPH increases throughout gestation; premature infants have initially a lower activity than full-term infants. It is present at maximum hydrolytic capacity at the time of birth in full-term infants and then declines at the time of weaning to about 10% of the neonatal value. Several genetically determined mechanisms are involved in this decline, acting both at the transcriptional and post-translational levels. Conversely, persistence of high lactase activity in adults seems to be caused by a single factor: an increased biosynthesis of lactase (i.e. increase of the transcription of the gene).

(b) Factors influencing lactose digestion

Lactose digestion is influenced by three factors: the digestion/absorption surface, the quantity of lactase, and the speed of the gastrointestinal transit.

Digestion/absorption surface
The digestion/absorption surface can be decreased during acute or chronic intestinal diseases.

Regulation of lactase activity
The intestinal lactase activity cannot be induced by lactose, and the natural decline in lactase cannot be prevented by chronic lactose feeding (Sahi, 1994; Semenza and Auricchio, 1995). Pancreatic proteases decrease lactase activity, probably through digestion of the LPH molecule at the brush-border membrane (Semenza & Auricchio, 1995). Some bacterial proteases, which can be present in the small bowel during bacterial overgrowth syndromes or in animals who harbour more bacteria in the small intestine than man, have the same effect.

Gastrointestinal transit time
Gastrointestinal transit is an important factor modulating lactose digestion, especially in lactase-deficient subjects. The flow rate of lactose in the jejunum (where LPH reaches its highest level) is influenced by gastric emptying. Factors that delay gastric emptying enhance lactose digestion. This explains why lactose from whole milk is better

digested than lactose from skim milk, and why lactose consumed together with cocoa or dietary fibre or a solid meal is better digested than lactose alone (see references in Marteau *et al.*, 1990)

4.2.2 Lactose maldigestion

Even when maintained at seemingly high levels, the LPH activity is low when compared to other saccharidases, and is the rate-limiting step of lactose assimilation.

(a) Adult-type hypolactasia

The most frequent cause of lactose maldigestion in adults is the 'adult-type hypolactasia'. In this case, the low lactase activity is due to the physiological decline of LPH with age (Flatz, 1995). Adult-type hypolactasia is also referred to as 'lactase restriction', 'lactase deficiency', 'lactase non-persistence' and 'primary lactose maldigestion/malabsorption'. Since the LPH level is usually not nil, these terms should be preferred to the term 'alactasia' which would imply a total lack of lactase activity. As mentioned above, no substrate-induced enzymatic adaptation was found with intestinal lactase. The link between weaning and the beginning of the decline in the intestinal lactase activity is thus not causative (Sahi, 1994; Flatz, 1995). Family studies with pedigree analysis, measurements of relative lactase activity and twin studies have shown that the lactase phenotype is genetically determined (Flatz, 1995). The observed segregation is satisfactorily explained by a Mendelian system with two alleles: LAC*R (lactase restriction gene(s)), and LAC*P (lactase persistence gene(s)), the latter being dominant over the recessive LAC*R. The LPH gene is located on chromosome 2.

The prevalence of adult-type hypolactasia varies between races and populations (Scrimshaw and Murray, 1988; Sahi, 1994; Flatz, 1995; Figure 4.1). Lactase restriction is predominant and often the ubiquitous lactase phenotype in the native populations of Australia and Oceania, east and Southeast Asia, tropical Africa, and the Americas. Predominance of the lactase-persistence phenotype is found in central and northern Europe (with a decreasing north to south gradient), and in nomadic populations of North Africa and Arabia (Flatz, 1995). Overall, 70–90% of adults of the world, and 30–50% of adults in Europe are 'lactase deficient'.

(b) Congenital lactase deficiency

Congenital lactase deficiency is an exceptional inborn error. No more than 40 cases have been reported (Semenza and Auricchio, 1995). It is

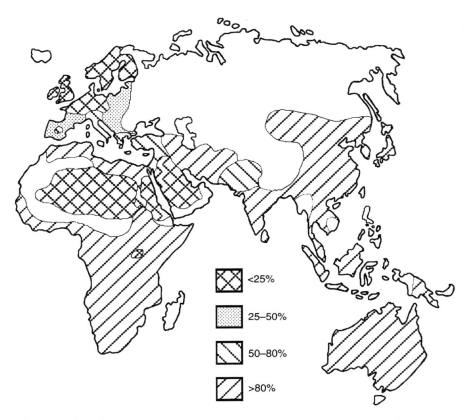

Fig. 4.1 Distribution of lactase phenotypes (shown as per cent of population with low lactase digestion capacity, see key) in the native populations of the Old World. Blank areas = insufficient data. Reproduced with permission from Flatz (1995).

characterized by a very low level of intestinal lactase without change of the other intestinal saccharidases. Symptoms, dominated by a watery acidic diarrhoea, occur in the newborn, and resolve during a lactose-free diet.

(c) Secondary lactose maldigestion

The terms 'secondary lactose maldigestion' or 'secondary hypolactasia' are used when lactose maldigestion is due to a disease affecting either the enterocytes, the absorption surface or the gastrointestinal motility. Main causes are acute or chronic viral, bacterial or parasitic enteritis (e.g. rotavirus diarrhoea), coeliac disease, tropical sprue, Whipple disease, protein-energy malnutrition, cow-milk allergy, actinic enteritis, small intestine bacterial overgrowth, Crohn's disease and extensive resection of the small bowel (often referred to as 'short bowel syndrome') (Scrimshaw and Murray, 1988). The two main causes of secondary lactose maldigestion implying a too fast intestinal transit are gastrectomy and vagotomy.

4.2.3 Physiological consequences of lactose maldigestion

When lactose maldigesters ingest lactose, a part of it is not hydrolysed and thus not absorbed in the small bowel. The non-absorbed lactose exerts an osmotic effect which induces secretion of water and sodium into the lumen. This increases the volume of the chyme and accelerates the transit of small bowel contents (Rambaud, Bouhnik and Marteau, 1994). The absorption of the other nutrients may be affected as indicated by higher quantities of protein, calcium, magnesium and phosphate in the ileum of lactase-deficient subjects after ingestion of milk (Debongnie *et al.*, 1979).

The part of lactose reaching the colon can be fermented (usually extensively) by the colonic bacteria or remain intact (being then still osmotically active in the colon and responsible for osmotic diarrhoea). Several species of the colonic flora exhibit lactase activities (i.e. β-galactosidases). Fermentation leads to the production of lactate, short-chain fatty acids (SCFA, i.e. acetate, propionate and butyrate), and of gases: hydrogen (H_2), carbon dioxide (CO_2), and in some subjects methane (CH_4). The SCFAs are responsible for an acidification of the colonic contents, at least in the right colon, and for faeces when intestinal transit is fast. A large part of the SCFAs is rapidly absorbed together with sodium and water and this allows both a salvage of energy, and a decrease in the volume of the chyme (antidiarrhoeal effect). This 'colonic salvage' has been shown to be important in the energy balance of neonates, especially in premature infants. The part of SCFAs which is not absorbed is excreted in faeces. An excessive production/absorption of lactate can lead to lactic acidosis.

Fermentation of lactose by the colonic flora is, like that of the closely related sugar lactulose, an adaptive process. A chronic ingestion of lactulose or of lactose in lactose maldigesters (i.e. a chronic load of lactose to the colonic flora) leads to changes in the fermentation: faecal levels of β-galactosidase increase, colonic pH decreases and breath H_2 excretion decreases (Florent *et al.*, 1985). These changes are referred to as 'metabolic adaptation'. Florent *et al.* (1985) showed that lactulose consumption not only leads to a metabolic adaptation but also to a clinical adaptation characterized by a decrease in the intolerance symptoms. Despite the metabolic adaptation, and some optimistic reports, no relevant clinical adaptation has been shown to chronic lactose consumption in lactose-intolerant subjects. In the majority of subjects, the part of lactose reaching the colon is totally fermented and no sign of intolerance is noticed.

4.2.4   Lactose intolerance

Some subjects experience intolerance signs when ingesting lactose. The appearance of symptoms depends not only on the amount of lactose ingested, but also on several physiological factors.

(a) Who are lactose-intolerant subjects?

The tolerance to lactose depends on the degree of lactose maldigestion, the physiology of the colon and its flora, and the visceral sensitivity of the subject (Figure 4.2). As mentioned above, the degree of lactose maldigestion depends on the quantity of ingested lactose, the lactase activity, the rate of gastric emptying and the motor response of the small bowel to the osmotic load. The frequency of lactose intolerance increases with the dose of ingested lactose or milk. However, the range of tolerance for lactose is extremely wide among lactase-deficient subjects. In addition, subjects with lactose intolerance frequently quote symptoms during placebo periods (e.g. lactose-free periods), and some subjects claiming to be milk intolerant digest lactose extensively

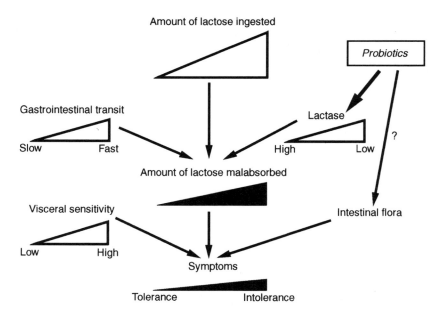

Fig. 4.2 Factors that influence the digestibility and tolerance of lactose.

(Suarez, Savaiano and Levitt, 1995). This underlines the need for control periods in clinical studies, and raises the question of the role of irritable bowel syndrome in lactose intolerance. A threshold for appearance of symptoms is very difficult to define for the whole population; nevertheless, only a minority of subjects have intolerance for doses below 3 g. Newcomer and McGill (1984) calculated on the basis of 14 studies that the mean prevalence of intolerance for one big glass of milk (i.e. 12 g of lactose) was 19% among lactose maldigesters (range 0–75%).

(b) Pathophysiology of the symptoms of intolerance

All symptoms of lactose intolerance do not have the same pathophysiology (Rambaud, Bouhnik and Marteau, 1994). Two main mechanisms are involved: osmotic effect and 'hyperfermentation'. Diarrhoea is due to an osmotic effect, largely because of the presence of the non-absorbed and non-fermented lactose in the chyme. The risk of diarrhoea increases when the capacity of the colonic flora to ferment carbohydrates decreases, for example during antibiotic treatment. Contrary to its beneficial action on diarrhoea, colonic fermentation of lactose is responsible for bloating, and increased gas production. Borborygmi and abdominal pain are probably due to both mechanisms, pain being also favoured in some subjects by increased intestinal sensitivity (Rambaud, Bouhnik and Marteau, 1994; Malagelada, 1995). The role of lactose intolerance in recurrent abdominal pain in children and in irritable bowel syndrome in adults has been debated. Although some overlap can exist between these situations, they are usually unrelated (Scrimshaw and Murray, 1988; Lynn and Friedman, 1993; Malagelada, 1995).

4.3   METHODS TO STUDY LACTOSE DIGESTION

Several methods have been developed to diagnose or quantify lactose maldigestion or hypolactasia (Brummer, Karibe and Stockbrügger, 1993).

4.3.1   Direct measurements

Assessment of intestinal LPH activity on biopsy samples of the jejunum was the first method used. This method is invasive and provides information on the lactase activity only at the site of sampling. The quantity of lactose maldigested in the small bowel can be directly measured in ileal fluid of subjects with an ileostomy or when using invasive intestinal intubation techniques in volunteers.

4.3.2   Indirect measurements

Indirect tests have been developed to study the metabolism of a lactose load by measuring one of its end products in breath, urine or blood. Their main advantage is that they are not invasive and can be routinely used. The sensitivity of these tests increases with the amount of lactose given, but even subjects with high lactase activity can exhibit some maldigestion of a high lactose load. It was initially decided to use a load consisting of 50 g of lactose in water or milk. However, there is a trend to decrease this dose to amounts closer to that consumed in everyday life, such as 20 g.

The breath hydrogen test is based on the determination of hydrogen from expired air after an oral lactose load. Hydrogen (H_2) production takes place only when a sugar reaches the colonic flora. Recent antibiotic treatment or laxatives can disturb the test. Subjects must avoid fermentable foods the evening before. Lactose is given orally to fasting subjects (who have less than 10 parts per million (ppm) molecules of H_2 in breath), and exhaled air is collected in airtight syringes every 30 min for H_2 measurement. The subjects have to refrain from eating, smoking and physical activity during the sampling period which lasts for 4 h for the usual test and 6–8 h when the whole fermentation of the lactose load has to be studied (in quantitative studies). A positive test is defined as a rise in H_2 concentration more than 20 ppm. When quantification of the lactose maldigestion is needed, the quantity of H_2 excreted after ingestion of lactose is compared to that excreted after ingestion of 10 g lactulose (which is totally maldigested in the small bowel). Breath tests assessing excretion of $^{13}CO_2$ after ingestion of ^{13}C-lactose, and using a ^{13}C-glucose test as reference can also be used (Brummer, Karibe and Stockbrügger, 1993).

The urinary galactose appearance, with or without additional ethanol to inhibit the hepatic transformation of galactose into glucose, is another simple and valid indirect test. Breath H_2 and galactose appearance tests have supplanted the test based on the assessment of the rise in blood glucose which is less sensitive. During lactose load tests, clinical tolerance can be assessed using questionnaires or visual analogue scales on the presence and severity of gastrointestinal symptoms such as abdominal bloating, flatulence, abdominal pain and loose stools.

4.3.3   Quantitative results

The methods generally used for quantifying maldigested lactose are breath hydrogen test with lactulose standard and intestinal intubation technique, which are both valid methods. The use of the former method should, however, be limited for group comparisons. Only a

few studies have been performed to quantify the maldigestion of lactose from different sources in lactose maldigesters. In studies using intestinal intubation techniques, the level of lactose maldigestion was 42–75% from a 12.5 g load of lactose in water (Bond and Levitt, 1976) and
0–50% from a 10 g load of lactose in unmodified milk (Debongnie *et al.*, 1979). All the studies, in which the maldigestion of lactose from different sources of dairy products has been measured, have consistently shown only a weak maldigestion of lactose from yoghurt or other semi-solid fermented milk products compared with that from unmodified milk (Kolars *et al.*, 1984; Marteau *et al.*, 1990; Arrigoni *et al.*, 1994; Vesa *et al.*, 1996).

4.4   DIGESTION AND TOLERANCE OF FERMENTED DAIRY PRODUCTS IN LACTOSE-INTOLERANT SUBJECTS

4.4.1   Yoghurt

(a) The yoghurt story

Gallagher, Molleson and Calwell (1974) were the first to demonstrate the better tolerance of lactose from fermented dairy product sources. They studied the tolerance of three diets in three lactase-deficient subjects. Two diets provided 50 g lactose: one as fermented dairy products (buttermilk, yoghurt and cottage cheese), the other as non-fermented dairy products (milk, ice cream and milk powder). The third diet was lactose free. The fermented milk diet was well tolerated whereas the consumption of non-fermented milk diet caused intolerance symptoms. Discussing potential explanations the authors suggested that 'the bacteria may continue to exert lactase activity in the intestinal tract after ingestion' or that the effect could be related to 'a delay in gastric emptying and intestinal motility'. Many studies (Table 4.1) have confirmed Gallagher's observation and support her conclusion that 'nutritionists and practising dieticians who instruct lactase-deficient patients on lactose-free diets should encourage these patients to test their tolerance to fermented dairy products'.

A second important step forward was reached in 1984 when teams in the universities of Minnesota and Oklahoma (Gilliland and Kim, 1984; Kolars *et al.* 1984; Savaiano *et al.*, 1984) reported a significant reduction of maldigestion of lactose from yoghurt using the newly developed breath H_2 test. Kolars *et al.* (1984) compared, in 10 lactase-deficient subjects, the digestion and tolerance of lactose in solution, in milk or in yoghurt. Ingestion of 18 g lactose in yoghurt resulted in about one-third of the breath H_2 excretion that was produced by a similar dose of lactose in water or milk (Figure 4.3). Diarrhoea or flatulence occurred

Lactose maldigestion

Table 4.1 Results of studies comparing the digestibility or tolerance of lactose from yoghurt (Y), heated yoghurt (HY), and milk (M) in lactose malabsorbers

Reference	Subjects	End points	Result
Gallagher, Molleson and Caldwell, 1974	L1	Symptoms	FDP > M
Gilliland and Kim, 1984	LM	H_2	Y > HY > M
Kolars et al., 1984	LM	H_2	Y > M
Savaiano et al., 1984	LM	H_2	Y > HY (P=0.08)
		Symptoms	Y=HY
McDonough et al. 1987		H_2	Y > HY > M
			Y=HY+lactase
Dewit et al., 1987	LM children	H_2	Y > HY > M
Martini et al., 1987	LM	H_2	Y > M
Wytock and DiPalma, 1988	LM	H_2	Y > M
Onwulata, Rao and Vankineni 1989	LM	H_2	Y > M
Lerebours et al., 1989	LM	H_2	Y > M
		Symptoms	Y=HY > M
Pochart et al., 1989a	LM	Duodenal sampling	Y=HY
Marteau et al., 1990	LM	H_2	Y > HY > M
		Ileal sampling	Y > HY
Boudraa et al., 1990	LI children	Symptoms	Y > M
Martini et al., 1991b	LM	H_2	Y > M
Varela Moreiras et al., 1992	LM aged people	H_2	Y > M
Kotz et al., 1994	LM	H_2	Y=M
			High lactase Y > M
Arrigoni et al., 1994	Short bowel syndrome	Jejunal sampling &H_2	Y > > M

LI, lactose-intolerant subjects; LM, lactose malabsorbers; H_2, breath hydrogen test; FDP, fermented dairy products.

in 80% of the subjects after milk ingestion, but in only 20% after ingestion of yoghurt. Three subjects were also intubated with a double lumen tube. Assay of duodenal aspirates showed negligible lactase activity before ingestion of yoghurt but appreciable activity for at least 1 h after ingestion of yoghurt. This supported the hypothesis that bacterial lactase could be available in the small bowel to digest alimentary lactose.

At the same time, Gilliland and Kim (1984) were the first to show that lactose digestion was better after ingestion of yoghurt with living bacteria than after ingestion of heated yoghurt. This was confirmed by Savaiano et al. (1984) who observed in nine lactase-deficient subjects that H_2 excretion tended to be lower (P = 0.08) after ingestion of yoghurt containing living bacteria ($3 \times 10^8 \, g^{-1}$) and lactase (0.64

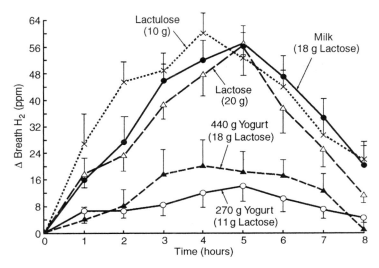

Fig. 4.3 Change in breath hydrogen after ingestion of lactose, milk yoghurt and lactulose. Values represent means ±1 SEM for 10 subjects. Reproduced with permission from Kolars *et al.* (1984).

units g^{-1}) than after ingestion of pasteurized yoghurt containing less living bacteria $(3.4 \times 10^6 \text{ g}^{-1})$ and less lactase (0.07 units g^{-1}). However, these subjects remained symptom-free with both products, whereas five and four of them exhibited symptoms after ingestion of sweet acidophilus milk and milk respectively.

These two series of experiments supported Gallagher's theory on the role of yoghurt lactase. They thus stimulated research in the fate and action of ingested yoghurt lactase, on the potential to vehiculate other enzymatic activities using other ingested micro-organisms (Marteau *et al.*, 1993), and on the practical consequences for dietary advice in lactase-deficient subjects. In addition, a series of studies was performed to confirm or quantify with other tests the better digestion of lactose from yoghurt, and to establish whether significant relevant differences would exist between brands of yoghurts, strains, and fermented products.

Pochart *et al.* (1989a) showed that faecal β-galactosidase activity increased in lactase-deficient subjects who regularly ingested heated yoghurt while it remained unchanged in subjects who ingested yoghurt with living bacteria. They confirmed indirectly that significant amounts of lactose reach the colon after ingestion of heated yoghurt but not yoghurt. This was also confirmed in a study using a direct measurement of lactose and carbohydrates in ileal fluid (Marteau *et al.*, 1990).

Dewit, Pochart and Desjeux (1988) suggested that not only lactose digestion, but also absorption of free fatty acids and factors acting on insulin release were enhanced with yoghurt when compared to milk. Martini, Kukielka and Savaiano (1991a) showed that yoghurt reduced lactose maldigestion regardless of whether or not it was consumed with a meal. In the same work, the authors addressed an important question: would yoghurt consumption help in the digestion of additional lactose from other sources? Six test meals with yoghurt and increasing amounts of added lactose were fed to 10 lactase-deficient subjects. Lactose digestion was assessed using the breath H_2 test. Results showed that yoghurt did not aid the digestion of additional lactose consumed simultaneously. The authors proposed two explanations for this: saturation of the yoghurt lactase or saturation of the lactose-permease system which transports the lactose into the bacterial cell. One might also conceive that the additional lactose is concentrated in a liquid phase of the meal which would be emptied earlier from the stomach than the semi-liquid phase of yoghurt.

The results of studies which compared digestion and tolerance of lactose from milk, yoghurt and heated yoghurt are indicated in Table 4.1. The majority of studies have been performed in healthy subjects with adult-type hypolactasia. The better digestion of lactose from yoghurt has also been observed in several other situations of primary or secondary hypolactasia, for example in elderly subjects (Varela-Moreiras *et al.*, 1992), infants with diarrhoea (Dewit *et al.*, 1987; Boudraa *et al.*, 1990) and patients with short bowel syndrome (Arrigoni *et al.*, 1994). Industrial companies have put effort into comparing commercial or prototype products. Although some differences have been reported between products, their relevance is often questionable. For example, Wytock and DiPalma (1988) entitled one of their papers 'All yoghurts are not created equal'. However, this title did not reflect properly the results of the study, since the statistical comparison of the breath H_2 excretion, and the clinical tolerance between the three tested brands of yoghurt did not show differences. Martini *et al.* (1991b) compared the ability of different strains and species of lactic acid bacteria to help lactose digestion *in vivo*. The seven yoghurts which were tested, although made with different mixtures of *Streptococcus salivarius* ssp. *thermophilus* and *Lactobacillus delbrueckii* ssp. *bulgaricus* and differing in their β-galactosidase activity (from 3.8 to 6.68 IU), exhibited the same improvement in lactose digestion.

(b) Quantitative aspects

Most studies on lactose digestibility from yoghurt are not quantitative but only provide a comparison of the digestibility of lactose from different sources. However, some studies have also provided quantitative data in subjects with adult-type hypolactasia either by using a control period with lactulose for breath tests or by direct assessment of undigested lactose. In the study of Kolars and coworkers (1984), maldigestion of lactose from milk was around 50%; from yoghurt it was only one-third of that (i.e. about 17%). Savaiano *et al.* (1984) and McDonough *et al.* (1987) showed that maldigestion of lactose from yoghurt was one-fifth of that of milk. We measured the digestion of lactose using the intestinal perfusion technique in eight lactase-deficient subjects who ingested 18 g of lactose in different fermented products. The amount of lactose that reached the terminal ileum was 1.74 g after ingestion of yoghurt and 2.85 g after ingestion of heated yoghurt (Marteau *et al.*, 1990). Although the maldigestion of lactose from heated yoghurt is nearly twice that of yoghurt, the absolute difference after ingestion of 18 g lactose (e.g. 400 g of yoghurt) is only about 1 g. In one study performed in 17 subjects with short bowel syndrome, the lactose maldigestion from milk was approximately 50%, and that from yoghurt 24% (Arrigoni *et al.*, 1994).

4.4.2   Other 'microbe-containing dairy products'

Results concerning the digestibility or tolerance of lactose from other fermented or non-fermented dairy products are less homogeneous than those obtained with yoghurt. The characteristics of the products, which are now known to independently influence lactose digestion (lactase content, bacteria, liquid or semi-solid state), are often insufficiently described, probably because they were still unknown at the time when the studies were performed. While the results on the digestibility of a product are not questionable, it is often impossible to draw reliable conclusions on the role of the probiotic micro-organisms contained in it. A few studies have been made with 'bifidus milks' or 'acidophilus milks' or milks to which lactic acid bacteria were added. Most of these products contain less lactase than yoghurt and are more liquid (Shah, Fedorak and Jelen, 1992).

(a) Unfermented and fermented acidophilus and bifidus milks

A poor digestion and tolerance of lactose from milks containing *Lactobacillus acidophilus* has been reported in several studies (Payne *et al.*, 1981; Newcomer *et al.*, 1983; McDonough *et al.*, 1987, Onwulata, Rao and Vankineni, 1989; Martini *et al.*, 1991b; Dehkordi *et al.*, 1995). The numbers of viable lactobacilli in these products were in most of the studies not mentioned or very low ($<10^7$ cfu ml^{-1}); in addition, the viability can be questioned when frozen concentrate starter cultures were used. Also the β-galactosidase activity and the physical state of the products were often not mentioned.

Hove, Nordgaard-Andersen and Portensen (1994) did not observe any modification of lactose digestibility between milk and the same milk plus capsules containing high numbers (3.5 10^9 cfu ml^{-1}) of *L. acidophilus* and *Bifidobacterium bifidum*. However, the survival of this preparation into the ileum was less than 1/100 and the vitality of the lyophilized bacteria can be questioned. Indeed, a much higher survival of bifidobacteria in the gastrointestinal tract has been observed when *L. acidophilus* and bifidobacteria were ingested in fermented dairy products (ref. in Marteau *et al.*, 1993).

An improvement of lactose digestion with acidophilus milks has been found in a few studies (Kim & Gilliland, 1983; Gaon *et al.*, 1995). Interestingly, McDonough *et al.* (1987) reported an improvement in lactose digestion only when the *L. acidophilus* cells were sonicated. This led to the hypothesis that in the presence of bile, the better resistance of *L. acidophilus* compared to yoghurt bacteria could limit the access of lactose to the bacterial cell where it is metabolized. Kim and Gilliland (1983) reported that the addition of high numbers of *L. acidophilus* to milk immediately prior to its ingestion (2.5 × 10^6 or 2.5 × 10^8 cfu ml^{-1}) improved lactose utilization. One may assume that such addition of bacterial cells at the last minute before ingestion would not influence the liquid state of the tested products.

We did not find differences in the digestibility between a fermented semi-solid milk product containing *L. acidophilus* and *Bifidobacterium* sp. and yoghurt bacteria with a lactase activity of 0.24 IU g^{-1}, and conventional yoghurt with a lactase activity of 0.8 IU g^{-1}. The maldigestion of lactose from these products varied between 18 and 21% (Vesa *et al.*, 1996). Both products were well tolerated. Feeding cultured buttermilk with *Lactococcus lactis* ssp. *lactis*, which contains only a phospho-β-galactosidase, did not improve lactose digestion in one study (Savaiano *et al.*, 1984). Martini *et al.* (1991b) compared lactose digestibility from low-fat milk, yoghurt and four milks fermented by a single strain of *S. salivarius* ssp. *thermophilus*, *L. delbrueckii* ssp. *bulgaricus*, *L. acidophilus* or *Bif. bifidum*. The physical state of the products was not mentioned. Their composition and the results are shown in Figure 4.4. Lactose digestion from the milks fermented by single strains was intermediate between that of milk (the 'worst') and yoghurt (the 'best').

The question of the degree of tolerance of fermented and non-fermented acidophilus or bifidus milks is thus still under debate. Based on the works of Vesa *et al.* (1996) and Shah, Fedorak and Jelen (1991, 1992), it can be proposed that the tolerance of liquid acidophilus products (as non-fermented milks usually are) is between that of yoghurt and milk, probably closer to milk. For semi-solid fermented milk products, a tolerance close to that of yoghurt and heated yoghurt can be expected.

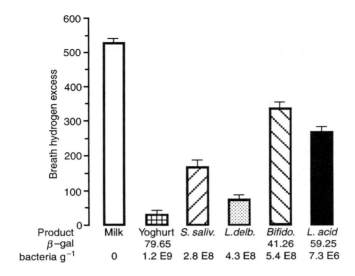

Fig. 4.4 Summation of the breath hydrogen concentrations produced every 30 min for an 8-h study period from meals of low-fat milk, yoghurt, and milk fermented by *Streptococcus salivarius* ssp. *thermophilus*, *Lactobacillus delbrueckii* ssp. *bulgaricus*, *Bifidobacterium bifidum* or *L. acidophilus*. Mean ±SE for 12 lactase-deficient subjects. Redrawn with permission from Martini *et al.* (1991b).

4.5   MECHANISMS FOR THE BETTER TOLERANCE AND DIGESTION OF LACTOSE FROM FERMENTED DAIRY PRODUCTS

4.5.1   Explanation for the better tolerance of lactose

Three hypotheses which are not mutually exclusive have been proposed to explain the well-established better tolerance of lactose from yoghurt.

(a) Less lactose

The first hypothesis raised to explain the better tolerance of yoghurt and fermented dairy products was their lower quantity of lactose due to the fermentation (Alm, 1982). However, many studies demonstrated that this explanation was not the only one. Yoghurts commercialized in many countries are supplemented with milk powder after fermentation and thus contain about the same quantity of lactose as milk ($40\,g\,l^{-1}$). Digestibility of lactose from these yoghurts remains superior to that of milk (Marteau *et al.*, 1990).

(b) Better clinical tolerance?

A change in intestinal sensitivity without change in lactose digestion could be imagined but has not been studied.

(c) Better lactose digestion

A better digestion of lactose from yoghurt and other fermented dairy products than from milk has clearly been shown, and this constitutes the main mechanism to explain the better tolerance (although the sensitivity hypothesis has not been properly studied and cannot be excluded).

4.5.2 Explanation for the better digestion of lactose

As in the case of better tolerance of lactose, three hypotheses, which do not exclude each other, have been proposed to explain the well-established better digestion of lactose from yoghurt.

(a) Lactase transport by the transiting micro-organisms

Arguments supporting an important role for microbial lactase
Several studies have suggested an important role of bacterial lactase. As mentioned above, digestion of lactose from heated yoghurt with no lactase activity is inferior to that from fresh yoghurt (Savaiano *et al.*, 1984; Dewit, Pochart and Desjeux, 1988; Lerebours *et al.*, 1989; Marteau *et al.*, 1990; Varela-Moreiras *et al.*, 1992). Ingestion of additional lactase isolated from yeasts and moulds in the form of capsules increases lactose digestion in lactase-deficient children or adults (Corazza *et al.*, 1992). Kotz *et al.* (1994) showed a significant improvement of lactose digestion by increasing the lactase activity of yoghurt five- to six-fold. Also in non-fermented milk, an increase of yoghurt bacteria from 10^7 to 10^8 cfu ml^{-1} improved lactose digestion in some studies (Lin, Savaiano and Harlander, 1991).

The passage and activity of the yoghurt lactase in the small bowel of humans consuming yoghurt has been demonstrated using intestinal intubation experiments. Kolars *et al.* (1984) detected yoghurt lactase in the duodenum for more than 1 h after yoghurt ingestion. Pochart *et al.* (1989b) confirmed this passage and showed that the yoghurt lactase was not active in the duodenum on lactose digestion probably because of the too acid pH. We observed that about one-fifth of the lactase ingested in yoghurt survived till the ileum, and was effective in the small bowel, allowing more digestion of lactose than what was observed with heated yoghurt (Marteau *et al.*, 1990; Figure 4.5).

An intact microbial cell structure is important for the survival of lactase in the stomach (Martini *et al.*, 1987). Marked heating and sonica-

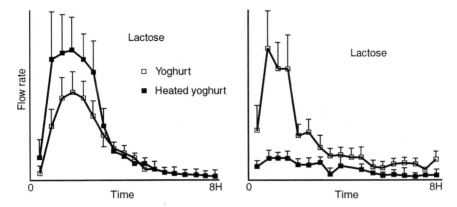

Fig. 4.5 Lactose and lactase flow rates through the ileum after ingestion of yoghurt or heated yoghurt. Values are means with SE for eight lactase-deficient subjects.

tion disrupt the cell structure of yoghurt bacteria and significantly elevate maldigestion while reducing the survival of microbial lactase. Yoghurt bacteria and lactase are destroyed as pH below 3.0 (Martini *et al.*, 1987; Pochart *et al.*, 1989b; Marteau *et al.*, 1993). Fortunately, yoghurt is an excellent buffer of acid due to its casein, lactate and calcium phosphate content, and the gastric pH remains above 2.7 for 3 h after yoghurt ingestion (Martini *et al.*, 1987).

In the small intestine, the pH rises in the distal duodenum and is optimal in the jejunum for bacterial lactase to be active. Bacterial lactase is an intracellular enzyme; for lactose to be digested, either it must enter the bacterial cell through a permease system or free lactase must be released through lysis of the bacteria. Yoghurt bacteria rapidly die in the small intestine because of a dose-related effect of bile acids which disrupt their cell structure (ref. in Marteau *et al.*, 1993). Bile increases lactase efficiency under *in vitro* conditions (Gilliland & Kim, 1984). The sensitivity of yoghurt bacteria to bile has been proposed as an advantage for lactose digestion, because it increases the permeability of the bacterial cell (Gilliland & Kim, 1984; McDonough *et al.*, 1987), and allows delivery of active lactase in the intestinal lumen. Additional evidence for the beneficial role of cell lysis has been provided by McDonough *et al.*, (1987) who showed that sonication of the comparatively bile-resistant *L. acidophilus* increased its efficacy for lactose digestion in humans. An influence of the permease of yoghurt bacteria on the effect on lactose digestion has been hypothesized to explain why bacteria with different intrinsic lactase activities have the same effect on lactose digestion (Martini *et al.*, 1991a, b). According to the above-mentioned arguments, bacteria or 'probiotics' containing low levels of lactase and being more resistant to bile, such as *L. acidophilus* or bifidobacteria, should be less efficient in helping lactose digestion than yoghurt bacteria.

Demonstration that microbial lactase is not all

Several studies fail to support the role of bacterial lactase. Martini and coworkers (1991b) and Vesa *et al.* (1996) did not find any difference in breath H_2 excretion and little difference in the tolerance between yoghurts or fermented milks with different levels of lactase activity. The symptoms of intolerance were equally alleviated with both yoghurt and heated yoghurt in several studies (Kolars *et al.*, 1984; Marteau *et al.*, 1990; Table 4.1). Furthermore, the absolute difference in the degree of maldigestion of lactose from yoghurt and heated yoghurt, when measured directly by sampling the intestinal contents or assessed indirectly using breath tests and lactulose as a standard, is small (i.e. about 1 g after ingestion of 18 g of lactose in 400 g of yoghurt) (Kolars *et al.*, 1984; Marteau *et al.*, 1990).

(b) Slower gastrointestinal transit

Slowing down of the intestinal transit of the lactose-containing phase, for example through a delayed gastric emptying, allows a better use of the endogenous LPH and enhances lactose digestion. Several studies have shown, using the breath H_2 test, that the orocaecal transit time of yoghurt is longer than of milk in lactase-deficient subjects (Savaiano *et al.*, 1984; Dewit, Pochart and Desjeux, 1988; Wytock and DiPalma, 1988; Lerebours *et al.*, 1989; Marteau *et al.*, 1990). It could be argued that the slower intestinal transit of yoghurt is the consequence of the better digestion of lactose, and not one of its causes.

In order to answer this question, we measured the gastric emptying and jejunal passage of the liquid phase of yoghurt and of milk using a direct intubation technique in lactase-persistent subjects (Marteau *et al.*, 1991). The gastric emptying half-time was slower for yoghurt (59 min ± 16) than for milk (26 min ± 5) and also the half-time for the passage of a meal marker at the Treitz ligament was slower for yoghurt (67 min ± 17) than for milk (47 min ± 18). Several mechanisms could explain the delayed gastric emptying of lactose in yoghurt. The composition of food, including caloric value, fat content, proteins, and pH, influences gastric emptying. A delayed gastric emptying of the lactose-containing phase of yoghurt due to its semi-solid state has also been hypothesized (Shah, Fedorak and Jelen, 1992; Vesa *et al.*, 1996). Shah, Fedorak and Jelen (1992) showed that solid pasteurized quark with no lactase activity was better digested and tolerated than liquid quark whey which contained the same amount of lactose and viable yoghurt organisms. More studies are needed in this field.

(c) Stimulation of the endogenous lactase

Stimulation of the endogenous intestinal lactase activity by ingested yoghurt bacteria has been proposed by Besnier *et al.* (1983). These authors observed that the intestinal mucosal lactase activity in mice increased after ingestion of a diet supplemented with yoghurt. Collington, Parker and Armstrong (1990) reported that a probiotic medication consisting of a mixture of *L. plantarum, L. acidophilus, L. casei* and *Enterococcus faecium* slightly increased lactase, sucrase and tripeptidase activities in the small intestine of piglets before weaning but had no effect on dipeptidase activity. Some modulation of the endogenous flora was suspected (i.e. a real probiotic effect) since antibiotics exhibited the same kind of effect. Lerebours *et al.* (1989) reported that ingestion of 125 g fresh yoghurt containing viable *L. delbrueckii* ssp. *bulgaricus* and *S. salivarius* ssp. *thermophilus* three times a day for 8 days did not significantly modify the duodenal lactase activity in 16 subjects with adult-type hypolactasia. Although this important study does not rule out the possibility of yoghurt micro-organisms stimulating lactase activity more distally, it does not support Besnier's hypothesis.

Interesting results were obtained in this field with the yeast *Saccharomyces boulardii*. Buts *et al.* (1986) administered orally 1 g day^{-1} of freeze-dried viable *Sac. boulardii* to seven healthy human volunteers for 2 weeks. Intestinal enzymatic activities were followed on sequential jejunal biopsies. A marked increase in lactase, sucrase and maltase activities was observed during the yeast administration whereas mucosal protein content and histological morphology remained unchanged. In the absence of yeast intrinsic lactase activity, the mechanism for the increased enzyme activities might be the stimulation of their synthesis at translational level, interference with their proteolytic decay or release from microvilli. In a following study, they showed that the increase in sucrase and maltase (lactase was not studied) induced in rats by *Sac. boulardii* was mediated by the endoluminal release of the polyamines spermine and spermidine (Buts, De Keyser and De Raedemaeker, 1994). Whether ingestion of *Sac. boulardii* would increase LPH activity in lactase-deficient subjects and would enhance lactose digestion has not been studied until now.

4.6 CLINICAL APPLICATIONS OF PROBIOTICS OR FERMENTED MILKS IN THE FIELD OF LACTOSE MALDIGESTION OR INTOLERANCE

As reported above, many studies have shown the better digestibility of lactose from yoghurt compared to milk in primary or secondary lactose maldigestion, and rational explanations have been discovered for this effect. The tolerance of other fermented dairy products is more debatable, and no information exists concerning beneficial effects of probiotics on lactose intolerance other than those contained in dairy products. In addition, one study shows improvement of symptoms in lactose maldigesters after a long-term administration of clinical preparation of *Bifidobacterium longum* and *Enterococcus durans* (Lehtniemi & Lindholm, 1995).

In a subject with 'primary lactose intolerance' due to 'adult-type hypolactasia', a lactose-free diet can mostly be avoided when the consumption of fermented dairy products is advised, especially that of yoghurt instead of milk.

Secondary lactose maldigestion is frequent, and it is therefore possible that enhancement of lactose digestion might participate in the beneficial antidiarrhoeal effects observed with 'probiotics' in different clinical situations (Marteau *et al.*, 1993). This is especially true and more relevant in children who consume larger amounts of lactose than adults. Replacing milk with yoghurt has proved to be beneficial in situations of persistent or chronic diarrhoea in children (Dewit *et al.*, 1987; Boudraa *et al.*, 1990), and this is of great practical importance. Several randomized studies have demonstrated a relevant therapeutic effect of *L. casei* strain GG during acute gastroenteritis in children (Isolauri, Juntunen and Rautanen, 1991; Majamaa *et al.*, 1995). Some studies have suggested the efficiency of *Bifidobacterium* preparations on similar cases of persistent paediatric diarrhoea (ref. in Marteau *et al.*, 1993). We observed in a series of subjects with short bowel syndrome that lactose maldigestion was constant, that about 50% of the lactose from milk was maldigested, and that a diet providing 20 g of lactose per day in the form of yoghurt, cheese or other solids was as well tolerated as a lactose-free diet (Arrigoni *et al.*, 1994). In other words, a lactose-free diet can be avoided in these subjects by using non-milk sources of lactose, especially yoghurt which is better digested.

4.7 COMMENTS AND CONCLUSIONS

Several studies have shown 'statistically significant' differences in lactose digestion among different fermented or non-fermented dairy products containing 'probiotics'. Several questions are raised:

• Are these differences relevant for the consumer?
• To what extent are the probiotics involved in the differences of lactose digestion?
• Are the effects on lactose digestion really 'probiotic effects'?

4.7.1   Are the differences in lactose digestibility between fermented dairy products relevant for the consumer?

This question is difficult to answer for two reasons:

- Most of the available studies were performed under unusual conditions favouring maldigestion (i.e. after ingestion of large quantities of dairy products in the fasting state). In other words, researchers select conditions to demonstrate 'differences' between products, but these conditions are sometimes far from reality. Extrapolation to normal conditions is difficult.
- Differences in the digestion between products often concern only biological, not clinical parameters. To our view, any difference, even small, is relevant if it concerns clinical signs of intolerance, but if it concerns only a digestion test, the relevance is more questionable. The significance of the partial loss of digestibility of lactose during pasteurization of dairy products initially containing living bacteria is questionable. Pasteurized products have a longer shelf-life and would be easier to handle, especially in poor countries. More clinical data are needed in this field, which has important economic and health consequences.

4.7.2   To what extent are the probiotics involved in the differences of lactose digestion?

Fermented and non-fermented dairy products differ considerably from each other not only in their bacteriological composition and lactase activity, but also in biochemical composition and physical properties. We now know that these characteristics can independently influence the digestibility of lactose. It is not possible to draw conclusions on the role of probiotics from studies comparing digestibility of dairy products without taking into account all these parameters, and many published studies are thus difficult to interpret.

4.7.3   Are the effects on lactose digestion really probiotic effects?

The mechanisms explaining the action of 'probiotics' on lactose digestion so far do not involve transformation or improvement of the endogenous flora. Therefore we think that these effects should be classified as direct effects of the transiting micro-organisms (in the lumen when bacterial lactase is involved or in the gut wall when stimulation of intestinal lactase is proposed), rather than true probiotic effects (Marteau *et al.*, 1993).

4.8 REFERENCES

Alm, L. (1982) Effect of fermentation on lactose, glucose, and galactose content in milk and suitability of fermented milk products for lactose intolerant individuals. *J. Dairy Sci.*, **65**, 346–52.

Arrigoni, E., Marteau, P., Briet, F. *et al.* (1994) Tolerance and absorption of lactose from milk and yogurt during short bowel syndrome in humans. *Am. J. Clin. Nutr.*, **60**, 926–9.

Besnier, M.O., Bourlioux, P., Fourniat, J. *et al.* (1983) Influence de l' ingestion de yogurt sur l'activité lactasique intestinale chez des souris axéniques ou holoxéniques. *Ann. Microbiol.*, **134**, 219–30.

Bond, J.H. and Lewitt, M.D. (1976) Quantitative measurement of lactose absorption. *Gastroenterology*, **70**, 1058–62.

Boudraa, G., Touhami, M., Pochart P. *et al.* (1990) Effect of feeding yogurt versus milk in children with persistent diarrhea. *J. Pediat. Gastroenterol. Nutr.*, **11**, 509–12.

Brummer, R.J.M., Karibe, M. and Stockbrügger R.W. (1993) Lactose malabsorption: optimalization of investigational methods. *Scand. J. Gastroenterol.*, **28** (Suppl. 200), 65–9.

Buts, J.P., de Keyser, N. and de Raedemaeker L. (1994) *Saccharomyces boulardii* enhances rat intestinal enzyme expression by endoluminal release of polyamines. *Pediat. Res.*, **36**, 522–7.

Buts, J.P., Bernasconi, P., Van Craynest, M.P. *et al.* (1986) Response of human and rat small intestinal mucosa to oral administration of *Saccharomyces boulardii*. *Pediat. Res.*, **20**, 192–6.

Collington, G.K., Parker, D.S. and Armstrong, D.G. (1990) The influence of inclusion of either an antibiotic or a probiotic in the diet on the development of digestive enzyme activity in the pig. *Br. J. Nutr.*, **64**, 59–70.

Corazza, G.R., Benati, G., Sorge, M. *et al.* (1992) β-galactosidase from *Aspergillus niger* in adult lactose malabsorption: a double-blind crossover study. *Aliment. Pharmacol. Therapy*, **6**, 61–6.

Debongnie, J.C., Newcomer, A.D., McGill, D.B. *et al.* (1979) Absorption of nutrients in lactase deficiency. *Dig. Dis. Sci.*, **24**, 225–31.

Dehkordi, N., Rao, D.R., Warren, A.P. *et al.* (1995) Lactose malabsorption as influenced by chocolate milk, skim milk, sucrose, whole milk, and lactic cultures. *J. Am. Dietetic Ass.*, **95**, 484–6.

Dewit, O., Pochart, P. and Desjeux, J.F. (1988) Breath hydrogen concentration and plasma glucose, insulin and free fatty acid levels after lactose, milk, fresh and heated yogurt ingestion by healthy young adults with or without lactose malabsorption. *Nutrition*, **4**, 1–5.

Dewit, O., Boudraa, G., Touhami, M. *et al.* (1987) Breath hydrogen test and stools characteristics after ingestion of milk and yogurt in malnourished children with chronic diarrhoea and lactase deficiency. *J. Trop. Pediat.*, **33**, 177–80.

Flatz, G. (1995) The genetic polymorphism of intestinal lactase activity in adult humans, in *The Metabolic and Molecular Basis of Inherited Disease*, 7th edn (eds C.R. Scriver, A.L. Beaudet, W.S. Sly *et al.*), McGraw-Hill, London, pp. 4441–50.

Florent, C., Flourié, B., Leblond, A. *et al.* (1985) Influence of chronic lactulose ingestion on the colonic metabolism of lactulose in man (an *in vivo* study). *J. Clin. Invest.*, **75**, 608–13.

Gallagher, C.R., Molleson, A.L. and Caldwell, J.H. (1974) Lactose intolerance and fermented dairy products. *J. Am. Dietetic Ass.*, **65**, 418–19.

Gaon, D., Doweck, Y., Zavaglia, A.G. *et al.* (1995) Lactose digestion by milk fermented with human strains of *Lactobacillus acidophilus* and *Lactobacillus casei*. *Medicina*, **55**, 237–42.

Gilliland, S.E. and Kim, H.S. (1984) Effect of viable starter culture bacteria in yogurt on lactose utilization in humans. *J. Dairy Sci.*, **67**, 1–6.

Hove, H., Nordgaard-Andersen, I. and Portensen, P.B. (1994) Effect of lactic acid bacteria on the intestinal production of lactate and short-chain fatty acids, and the absorption of lactose. *Am. J. Clin. Nutr.*, **59**, 74–9.

Isolauri, E., Juntunen, M. and Rautanen, T. (1991) A human *Lactobacillus* strain (*Lactobacillus casei* sp GG) promotes recovery from acute diarrhea in children. *Pediatrics*, **88**, 90–7.

Kim, H.S. and Gilliland, S.E. (1983) *Lactobacillus acidophilus* as a dietary adjunct for milk to aid lactose digestion in humans. *J. Dairy Sci.*, **66**, 959–66.

Kolars, J.C., Levitt, M.D., Aouji, M. *et al.*, (1984) Yogurt: an autodigesting source of lactose. *N. Engl. J. Med.*, **310**, 1–3.

Kotz, C.A., Furne, J.K., Savaiano, D.A. *et al.* (1994) Factors affecting the ability of a high β-galactosidase yogurt to enhance lactose absorption. *J. Dairy Sci.*, **77**, 3538–44.

Lehtniemi, A. and Lindholm, R. (1995) Laktobasillivalmisteen vaikutus laktoosi-intoleranssin oireisiin (in Finnish). *Duodecim*, **111**, 1027–31.

Lerebours, E., N'Djitoyap Ndam, C., Lavoine, A. *et al.* (1989) Yogurt and fermented-then-pasteurized milk: effects of short-term and long-term ingestion on lactose absorption and mucosal lactase activity in lactase-deficient subjects. *Am. J. Clin. Nutr.*, **49**, 823–7.

Lin, M., Savaiano, D. and Harlander, S. (1991) Influence of nonfermented dairy products containing bacterial starter cultures on lactose maldigestion in humans. *J. Dairy Sci.*, **4**, 87–95.

Lynn, R.B. and Friedman, L.S. (1993) Irritable bowel syndrome. *N. Engl. J. Med.*, **329**, 1940–5.

Majamaa, H., Isolauri, E., Saxelin, M. *et al.* (1995) Lactic acid bacteria in the treatment of acute rotavirus gastroenteritis. *J. Pediat. Gastroenterol. Nutr.*, **20**, 333–8.

Malagelada, J.R. (1995) Lactose intolerance. *N. Engl. J. Med.*, **333**, 53–4.

Marteau, P., Flourié B., Pochart, P. *et al.* (1990) Effect of the microbial lactase (EC 3. 2. 1. 23) activity in yoghurt on the intestinal absorption of lactose: an *in vivo* study in lactase-deficient humans. *Br. J. Nutr.*, **64**, 71–9.

Marteau, P., Pochart, P., Mahé, S. *et al.* (1991) Gastric emptying but not orocecal transit time differs between milk and yoghurt in lactose digesters. *Gastroenterology*, **100**, A353 (abstr).

Marteau, P., Pochart, P., Bouhnik, Y. *et al.* (1993) Fate and effects of some transiting microorganisms in the human gastrointestinal tract. *Wd. Rev. Nutr. Dietetics*, **74**, 1–21.

Martini, M.C., Kukielka, D. and Savaiano, D.A. (1991a) Lactose digestion from yogurt: influence of meal and additional lactose. *Am. J. Clin. Nutr.*, **53**, 1253–8.

Martini, M.C., Bollweg, G.L., Levitt, M.D. *et al.* (1987) Lactose digestion by yogurt β-galactosidase: influence of pH and microbial cell integrity. *Am. J. Clin. Nutr.*, **45**, 432–6.

Martini, M.C., Lerebours, E.C., Lin, W. *et al.* (1991b) Strains and species of lactic aid bacteria in fermented milks (yogurts): effect on *in vivo* lactose digestion. *Am. J. Clin. Nutr.*, **54**, 1041–6.

McDonough, F.E., Hitchins, A.D., Wong, N.P. *et al.* (1987) Modification of sweet acidophilus milk to improve utilization by lactose-intolerant persons. *Am. J. Clin. Nutr.*, **45**, 570–4.

Newcomer, A.D. and McGill, D.B. (1984) Clinical importance of lactase deficiency. *N. Engl. J. Med.*, **310**, 42–3.

Newcomer, A.D., Park, H.S., O'Brien, P.C. *et al.* (1983) Response of patients with irritable bowel syndrome and lactase deficiency using unfermented acidophilus milk. *Am. J. Clin. Nutr.*, **38**, 257–63.

Onwulata, C.I., Rao, D.R. and Vankineni, P. (1989) Relative efficiency of yogurt, sweet acidophilus milk, hydrolyzed-lactose milk, and a commercial lactase tablet in alleviating lactose maldigestion. *Am. J. Clin. Nutr.*, **49**, 1233–7.

Payne, D.L., Welsh, S.D., Manion, C.V. *et al.* (1981) Effectiveness of milk products in dietary management of lactose malabsorption. *Am. J. Clin. Nutr.*, **34**, 2711–15.

Pochart, P., Bisetti, N. Bourlioux, P. *et al.* (1989a) Effect of daily consumption of fresh or pasteurized yogurt on intestinal lactose utilisation in lactose malabsorbers. *Microecology and Therapy*, **18**, 105–10.

Pochart, P., Dewit, O., Desjeux, J.F. *et al.* (1989b) Viable starter culture, β-galactosidase activity and lactose in the duodenum; after yogurt ingestion in lactase-deficient humans. *Am. J. Clin. Nutr.*, **49**, 828–31.

Rambaud, J.C., Bouhnik, Y. and Marteau, P. (1994) Dairy products and intestinal flora, in *Proceedings of the First World Congress of Dairy Products in Human Health and Nutrition*, Madrid, 7–10 June 1993 (eds M. Serrano Rios, A. Sastre, M.A. Perez Juez *et al.*), Balkema, Rotterdam, pp. 389–99.

Sahi, T. (1994) Genetics and epidemiology of adult-type hypolactasia. *Scand. J. Gastroenterol.*, **29**, (Suppl. 202), 7–20.

Savaiano, D.A., Abou ElAnouar, A., Smith, D.E. *et al.* (1984) Lactose malabsorption from yogurt, pasteurized yogurt, sweet acidophilus milk, and cultured milk in lactase-deficient individuals. *Am. J. Clin. Nutr.*, **40**, 1219–23.

Scrimshaw, N.S. and Murray, E.B. (1988) The acceptability of milk and milk products in populations with a high prevalence of lactose intolerence. *Am. J. Clin. Nutr.*, **48**, (Suppl.), 1083–159.

Semenza, G. and Auricchio, S. (1995) Small-intestinal disaccharidases, in *The Metabolic and Molecular Basis of Inherited Disease*, 7th edn (eds C.R. Scriver, A.L. Beaudet, W.S. Sly *et al.*), McGraw-Hill, London, pp. 4451–79.

Shah, N.P., Fedorak, R.N. and Jelen, P.J. (1992) Food consistency effects of quarg in lactose malabsorption. *Int. Dairy J.*, **2**, 257–69.

Suarez, F.L., Savaiano, D.A. and Levitt, M.D. (1995) A comparison of symptoms after the consumption of milk or lactose-hydrolyzed milk by people with self-reported severe lactose intolerance. *N. Engl. J. Med.*, **333**, 1–4.

Varela-Moreiras, G., Antoine, J.M., Ruiz-Roso, B. *et al.* (1992) Effects of yogurt and fermented-then-pasteurized milk on lactose absorption in an institutionalized elderly group. *J. Am. Coll. Nutr.*, **11**, 168–71.

Vesa, T.H., Marteau, P., Zidi, S. *et al.* (1996) Digestion and tolerance of lactose from yoghurt and different semi-solid fermented dairy products containing *Lactobacillus acidophilus* and bifidobacteria in lactose maldigesters – Is bacterial lactase so important? *Eur. J. Clin. Nutr.*, **50**, 730–3.

Wytock, D.H. and DiPalma, J.A. (1988) All yogurts are not created equal. *Am. J. Clin. Nutr.*, **47**, 454–7.

CHAPTER 5

Antimutagenic and antitumour activities of lactic acid bacteria

A. Hosono, H. Kitazawa and T. Yamaguchi

5.1 INTRODUCTION

Epidemiological studies suggest that around one-third of all human cancers are related to the diet (Doll and Peto, 1981). Many studies on the presence of natural mutagens, antimutagens and cancer-modulating compounds as inherent factors of our food have been published in the last decade (Ahmed, 1992; Norred and Voss, 1994). At one time it was the opinion in cancer research that there was no relationship between carcinogens and mutagens, but reports that 4-nitroquinoline-N-oxide (4NQO), a carcinogen, also had the property of inducing spontaneous mutation in bacteria, and that N-methyl-N'-nitro-nitrosoguanidine (MNNG), a known mutagen, also induced stomach cancer, led inexorably to the conclusion that there is a close relationship between the two, and today it is clear that 85% of carcinogens are mutagenic (Sugimura, 1982). This indicates that there may be a steady increase in substances combining mutagenicity and carcinogenicity in our surroundings.

Lactic acid bacteria and their fermented food products are thought to confer a variety of important nutritional and therapeutic benefits on consumers, including antimutagenic and anticarcinogenic activity. At the beginning of this century, Elie Metchnikoff proposed that the lactic acid bacteria contained in soured milk would be of value in promoting the health of the human race if this form of fermented milk was brought into general use (Metchnikoff, 1908). During the past decade,

Probiotics 2: Applications and practical aspects.
Edited by R. Fuller.
Published in 1997 by Chapman & Hall.   ISBN 0 412 73610 1

several papers have reviewed the evidence for the suppression of carcinogenesis by lactic acid bacteria (Friend and Shahani, 1984; Fernandes, Shahani and Amer, 1987; Hosono, 1988; Gilliland, 1990; Adachi, 1992; Ballongue, 1993). Based on research reported during the past 10 years concerning antimutagenic and anticarcinogenic activity in fermented milk, Nadathur, Gould and Bakalinsky (1994) classified the findings into 5 groups:

- Antimutagenicity, which has been detected against a range of mutagens and promutagens in various test systems on microbial and mammalian cells.
- Consumption of fermented milk which inhibited growth of certain types of tumours in mice and rats.
- Oral supplements of *Lactobacillus acidophilus* in humans, which reduced activities of faecal bacterial enzymes (β-glucuronidase, nitrosoreductase and azoreductase) that are involved in procarcinogen activation and reduced excretion of mutagens in faeces and urine.
- Stimulation of the immune system of humans and other animals by dietary lactobacilli.
- Epidemiological evidence, which indicates a negative correlation between the incidence of certain cancers and consumption of fermented milk products.

This chapter reviews recent scientific data regarding both antimutagenic and antitumour activity of lactic acid bacteria.

5.2   ANTIMUTAGENIC ACTIVITY OF LACTIC ACID BACTERIA

Many substances have been discovered in food (and its components) which have the property of weakening the action of mutagens or carcinogens (Hosono, 1992). It is clear that mechanisms such as detoxification and DNA repair exist which protect against the effects of many classes of naturally occurring and synthetic mutagens. In fact, the term 'antimutagen' was used originally to describe those agents that reduce the frequency or the rate of spontaneous or induced mutation independent of the mechanisms involved (Novick and Szilar, 1952). It was later that distinction was made between antimutagenic agents acting outside the cell (desmutagens) and functioning inside (bioantimutagens) based on mechanism of action (Kada, Inoue and Namiki, 1981).

5.2.1   Nutritional and therapeutic benefits of milk fermented with lactic acid bacteria

Most of the indigenous communities colonize habitats in the large bowel. That region of the tract harbours over 10^{14} bacterial cells. These cells make up one half of the intestinal content. They can express a great variety of biochemical activities. Those activities and the bacterial populations expressing them are regulated in climax communities by a complex of factors that are incompletely understood. Such a complex also governs the successional pattern by which the microflora develops in infants. Some of the factors derive from the human host and its diet and environment. Others derive from the micro-organisms themselves. Some have a general influence, impacting on the entire microflora. Others are specific, influencing only certain of the communities (Savage, 1995). Among the intestinal micro-organisms, there are several potential health or nutritional benefits possible from some species of lactic acid bacteria. Among these are improved nutritional value of food, control of intestinal infections, improved digestion of lactose, control of some types of cancer and control of serum cholesterol levels. Some potential benefits may result from growth and action of the bacteria during the manufacture of cultured foods (Gilliland, 1990).

Metchnikoff (1908) was perhaps the first researcher to propose that fermented dairy products have beneficial properties. During the past two decades, there has been renewed interest in the study of the nutritional and therapeutic aspects of these products. While numerous researchers (Deeth, 1984; Gurr, 1987) have suggested that lactic cultures and cultured products provide several nutritional and therapeutic benefits to the consumer, there exist a few reports (Robinson, 1989) in which the benefits have been questioned. The major reasons for the differences in the above studies may be due to differences in the cultures used as well as the experimental procedures and systems employed in different studies (Friend and Shahani, 1984). The majority of the papers suggest that the potential benefit following the consumption of fermented dairy products containing viable lactic acid bacteria (Deeth, 1984; Friend and Shahani, 1984; Gilliland, 1989) is primarily attributable to the favourable alteration in gastrointestinal microecology. Consumption of fermented products containing viable organisms increases the count of lactic acid bacteria in the intestine and decreases the coliform count as observed by faecal analysis (Ayebo, Shahani and Dam, 1981; Ayebo *et al.*, 1982).

If pathogenic bacteria such as *Escherichia coli*, *Mycobacterium tuberculosis* or salmonellae are mixed with fermented milk, lactic milk or lactic milk drinks, the numbers of bacteria decrease during storage. Even if fermented milk or lactic milk drinks are neutralized with NaOH, the speed of multiplication of *E. coli* is slower than in an unfermented mix with the same milk solids content. These facts indicate that organic acids in fermented milks and lactic milk drinks decrease or kill harmful bacteria and as a result suppress the production of harmful substances such as amines, phenols, indole, scatol and hydrogen sulphide (Yukuchi, Goto and Okonogi, 1992).

On the other hand, therapeutic benefits associated with lactic acid bacteria include prophylaxis against some types of intestinal infection (Fernandes *et al.*, 1987), increased tolerance to lactose-containing foods (Savaiano and Levitt, 1987) and possible prevention of cancer initiation (Fernandes, Shahani and Amer, 1987). *In vitro* studies have indicated that lactic acid bacteria, including *Bifidobacterium* sp., inhibit the growth of some pathogens (Gilliland and Speck, 1977; Gibson and Wang, 1994; Gibson *et al.*, 1995). Although lactic acid bacteria may inhibit the growth of other bacteria by the production of organic acids, hydrogen peroxide, diacetyl, β-hydroxy-propionialdehyde, etc., bacteriocins have become of great interest lately. Bacteriocins are peptides or proteins that show a bactericidal mode of action against bacteria that are usually closely related to the producer culture (Klaenhammer, 1993). Toba, Yoshioka and Itoh (1991a) reported that *L. acidophilus* LAPT 1060, isolated from infant faeces, produced an antimicrobial agent effective against six strains of *L. delbrueckii* ssp. *bulgaricus* and six strains of *L. helveticus*. The agent was found to be a bacteriocin, and designated acidophilucin A. *L. delbrueckii* ssp. *lactis* JCM 1106 and 1107 produced bacteriocins (lacticin A and B) that were active against three subspecies of *L. delbrueckii* (Toba, Yoshioka and Itoh, 1991b). *L. gasseri* isolated from infant faeces also produces a bacteriocin (gassericin A). According to these authors, the high incidence of bacteriocin production in faecal *L. gasseri* suggests the *in vivo* conjugational transfer of the determinants of bacteriocin production and the important role played by bacteriocins in the control of intestinal microflora ecosystems.

Antifungal properties of lactic acid bacteria have also been described. Suzuki, Nomura and Morichi (1991) isolated *Leuconostoc mesenteroides*, which possesses an antifungal activity, from cheese starter. Recently, Kanatani, Oshimura and Sano (1995) purified acidocin A, a bacteriocin produced by *L. acidophilus* TK 9201 (Table 5.1). The molecular mass was 6500 Da. The sequence of the first 16 amino acids of the N terminus was determined, and oligonucleotide probes based on this sequence were constructed to detect the acidocin A structural gene *acd*A. The probes hybridized to the 4.5-kb *Eco*RI fragment of a 45-kb plasmid, pLA9201, present in *L. acidophilus* TK92201, and the hybridizing region was further localized to the 0.9-kb K*pnl*-X*bal* fragment. Analysis of the nucleotide sequence of this fragment revealed that acidocin A was synthesized as an 81-amino-acid precursor including a 23-amino-acid N-terminal extension (Kanatani, Oshimura and Sano, 1995).

Beside bacteriocins, several antibiotic-like substances such as acidolin (Hamdan and Mikolajicik, 1974), lactocidin (Vincent, Veomett and Riley, 1959), and other broad-spectrum inhibitors (Mehata, Patel and Dave, 1983; Silva *et al.*, 1987) have been reported to be produced by species of lactobacilli. Certainly, these types of compounds may be very effective in helping to control growth of intestinal pathogens. An

Table 5.1 Purification of acidocin A. From Kanatani, Oshimura and Sano (1995)

Purification step	Vol (ml)	Total protein (mg)	Total activity (AU)	Specific activity (AU mg^{-1})	Yield (%)	Purification (fold)
Culture supernatant	2000	46000	1.1×10^6	24	100	1
Ammonium sulphate precipitation	60	4900	7.2×10^5	1.5×10^2	65	6
Carboxymethyl-cellulose	20	9.2	1.7×10^5	1.8×10^4	15	750
Chromatography						
Sep-Pak cartridge (C$_{18}$ column)	5	1.8	1.2×10^5	6.7×10^4	11	2790
Aquapore RP-300 (C$_8$ column)	1	1.5	1.1×10^5	7.3×10^4	10	3040

important point to remember is that the intensity of the antagonistic action produced by lactobacilli varies among strains of the organism (Gilliland and Speck, 1977). Thus it is important in conducting experiments to evaluate the potential for the lactobacilli in controlling intestinal pathogens that some screening be included to select a culture most likely to exert maximum inhibitory action (Gilliland, 1990).

The effectiveness of fermented milk and lactic milk drinks against lactose maldigestion is also well known. The usual symptoms associated with this problem include cramps, flatulence and diarrhoea following the consumption of milk products. Those persons having the problem normally avoid including milk products in their diet. Children and adolescents represent a major segment of the milk-drinking population, and reduction or elimination of milk and dairy products from the diet of children with lactose maldigestion may compromise their intake of protein and calcium (Saavedra and Perman, 1989). Currently available validated approaches include the use of microbially derived lactase (β-galactosidase) added to milk to prehydrolyse the lactose. Montes *et al.* (1995) have recently considered whether the magnitude of lactose maldigestion (as measured by breath H_2 excretion) and its associated symptoms were reduced for children fed a physiological amount of lactose as 250 ml of low fat milk inoculated with 10^8 cells of *L. acidophilus*, compared with those of children fed uninoculated milk containing an equivalent amount of lactose. Nine of 10 subjects who were symptomatic following ingestion of uninoculated milk experienced a reduction in symptoms following ingestion of milk inoculated with *L. acidophilus* (Figure 5.1).

The discovery by Mann and Spoerry (1974) that people among the Masai warriors who drank large amounts (8.33 li day^{-1}) of a yoghurt fermented with wild strains of *Lactobacillus* sp. had very low values for

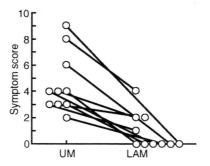

Fig. 5.1 Changes in symptom scores in 10 lactose-maldigesting children following ingestion of 250 ml of uninoculated low-fat milk (UM) or 250 ml of the same milk inoculated with *Lactobacillus acidophilus* (LAM). From Montes *et al.* (1995).

blood serum cholesterol opened up a new area of study. Gilliland, Nelson and Maxwell (1985) showed that *L. acidophilus*, when grown under proper conditions in the intestinal tract, will remove cholesterol from laboratory media. Harrison and Peat (1975) reported that cells of *L. acidophilus* added to infant formula decreased levels of serum cholesterol during the first 6–9 days of life. These observations suggest that the lactobacilli in the intestinal tract can and do influence serum cholesterol levels. Hepner *et al.* (1979) used 53 healthy men and women in two groups A and B, given cows' milk, normal food and yoghurt in different order, each for 4 weeks. Subjects in group A added three 240 ml cups of yoghurt to their normal diet daily for 4 weeks, then ate their normal diet only for 4 weeks and during a third 4-week period supplemented their diet with 720 ml of milk containing 2% butterfat. Subjects in group B followed a similar protocol, except that they supplemented their normal diet with three 240 ml cups of yoghurt during the third 4-week period. Figure 5.2 shows that blood cholesterol went down in all three groups when they were given yoghurt.

Contrary to these findings, Gilliland and Walker (1990) compared the assimilation of cholesterol by 12 commercially available cultures of *L. acidophilus* said to originate in human intestine and concluded that the cultures currently available commercially are probably not sufficiently active at assimilating cholesterol for best use in human diets to aid in the control of serum cholesterol levels. Using faecal isolates of *L. acidophilus* from human volunteers, Buck and Gilliland (1994) tested for bile salts and the ability to assimilate cholesterol during growth. In most cases, isolates of *L. acidophilus* from the same volunteer varied significantly in the amount of cholesterol assimilated, bile salt deconjugated and bile tolerance. The two cultures from each of nine volunteers that assimilated the most cholesterol were compared as a group to select the most active cultures. Significant variation in the ability to assimilate cholesterol was observed among these isolates from different volunteers. Eight of 17 isolates assimilated significantly.

5.2.2 Antimutagenic benefits of milk cultured with lactic acid bacteria

Lactic acid bacteria and their fermented food products are thought to confer a variety of important nutritional and therapeutic benefits on consumers, including antimutagenic and anticarcinogenic activity (Reddy, Shahani and Banerjee, 1973; Friend, Fermer and Shahani, 1982; Friend and Shahani, 1984; Fernandes, Shahani and Amer, 1987; Gilliland, 1990, 1991; O'Sullivan *et al.*, 1992). It is well known that carcinogenesis is initiated by mutation in animal cells under the influence of carcinogen. Evidence for the suppression of carcinogen mutations by lactic acid bacteria has been published (Hosono, 1988). The antimuta-

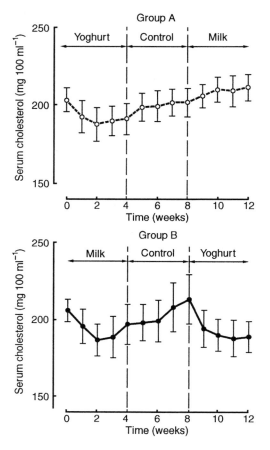

Fig. 5.2 Serum cholesterol, mg 100 ml^{-1} (mean ±SE) group A and group B subsjects. From Hepner *et al.* (1979).

genic acivity seemed to arise from some metabolites other than lactic acid.

Hosono, Kashina and Kada (1986) were the first to report that milk fermented with *L. delbrueckii* ssp. *bulgaricus* IFO 3533, *Lactococcus lactis* ssp. *lactis* IFO 12546 or *Enterococcus faecalis* exhibited antimutagenic activities against 4NQO, a typical mutagen, and the water extract of dog faeces, a faecal mutagen, *in vitro* using the *Escherichia coli* B/r WP2 trp⁻hcr⁻ strain. From the results of *in vitro* assay with streptomycin-dependent strains of *Salmonella typhimurium* TA 98 and TA 100, and AF-2, 4NQO and faecal mutagenic extracts from cats, monkeys, dogs and other animals, antimutagenic activities were detected (Hosono, Sagae and Tokita, 1986). The antimutagenic activities against the mutagenic chemicals increased with incubation time, but were lost rapidly at temperatures higher than 55°C. Nishioka *et al.* (1989) have reported that with respect to 49 strains of *Lactobacillus* sp., 18 of *Enterococcus* sp. and other strains tested, three strains of *Lactobacillus* sp., three strains of *Lactococcus* sp., two of *Streptococcus* sp. and one of *Enterococcus* sp. exhibited strong antimutagenic activity against 4NQO and 2-amino fluorine (AF-2) *in vivo* on *E. coli* trp⁻hcr⁻.

Antimutagenic activity of *Ent. faecalis* was strain specific. The extent of antimutagenic activity was similar to the desmutagenicity observed in cultured milk (Hosono, Sagae and Tokita, 1986), but that of some other strains was considerably lower. Since antimutagenic activity of some of the organisms was distinct from that of the culture medium, it seemed that antimutagenic activity of the organisms arises not only from metabolites produced by the organisms, but also from the bacterial cells. The higher antimutagenic activity of bacteria in milk compared with other media appeared to originate from antimutagenic activity of casein, especially β-casein, which was demonstrated as an active antimutagenic component in milk against the ethanol extract of pepper, a natural mutagen (Table 5.2) (Hosono, Shashikanth and Otani, 1989).

Bodana and Rao (1990) studied antimutagenic activity of acetone or ethylacetate extracts of skim milk fermented by *Streptococcus salivarius* ssp. *thermophilus*, *L. delbruekii* ssp. *bulgaricus* or a combination of both organisms using *Sal. typhimurium* (TA 98 and TA 100). Mutagens used were 4NQO (a direct-acting mutagen) and AF-2 (a mutagen requiring S9 activation). Extracts from all fermented milks showed significant ($P < 0.05$) dose response in suppressing the number of revertants caused by 4NQO and AF-2 in both tester strains, whereas extracts from unfermented milk had no effect. Extracts prepared from milk fermented by *L. delbrueckii* ssp. *bulgaricus* plus *S. salivarius* ssp. *thermophilus* showed significantly ($P < 0.05$) more antimutagenic activity than extracts prepared from milk fermented by *S. salivarius* ssp. *thermophilus*

Table 5.2 Antimutagenic activity of caseins and albumins. From Hosona, Shashikanth and Otani (1989)

Digestion	Revertants plate⁻¹
Control (H₂ O+pepper)	2012
Whole casein	844
α_{s1}-Casein	1085
β-Casein	609
κ-Casein	1741
Egg albumin	495
Bovine serum albumin	760

alone. Solvent (acetone vs. ethyl acetate) effect was not significant with 4NQO as mutagen. However, in the case of AF-2, acetone extracts showed significantly ($P < 0.05$) higher antimutagenic activity. The results of this and related studies strongly indicate that antimutagenic compounds are produced in milk during fermentation by *S. salivarius* ssp. *thermophilus* and *L. delbrueckii* ssp. *bulgaricus*.

Nadathur, Gould and Bakalinsky (1994) described antimutagenic activity extracted from fermented milk relative to mutagenesis induced by MNNG and 3, 2'-dimethyl-4-aminobiphenyl (DMAB). Reconstituted non-fat dry milk was fermented by *L. helveticus* CH65, *L. acidophilus* BG2F04, *S. salivarius* ssp. *thermophilus* CH3, *L. delbrueckii* ssp. *bulgaricus* 191R and by a mixture of the latter two organisms. The fermented milks were then freeze-dried, extracted in acetone, dissolved in dimethylsulphoxide, and assayed for antimutagenicity in the Ames test (*Sal. typhimurium* TA 100) against MNNG, and DMAB. Dose-dependent activity was significant against both mutagens in all extracts. Maximal inhibitory activity against DMAB and MNNG was 2- and 2.7-fold greater, respectively, than that exhibited by extracts of unfermented milk (Table 5.3). Compounds in the extracts of milk fermented by *L. delbrueckii* ssp. *bulgaricus* 191R were less soluble aqueous solutions than in dimethylsulphoxide. Adjustment of milk fermented by *L. delbrueckii* ssp. *bulgaricus* 191R to pH 3, 7.6 or 13 prior to freeze-drying and acetone extraction did not significantly alter the activity specific for DMAB. In contrast, compounds with activity specific for MNNG were less extractable at pH 7.6. The weak antimutagenicity of unfermented milk was not increased by addition of 2% L-lactic acid.

Kiyosawa, Matsuyama and Arai (1992) observed an antimutagenic effect of the aqueous and 80% methanol extracts of soymilk on MNNG mutagenesis toward *E. coli* B/r WP2 trp⁻. Addition of both extracts to the plate decreased the mutation frequency through the dose amount of 60 ml. However, there was a tendency for the mutation frequency to increase with the addition of more than 80 ml. The effect of the extract prepared from soybeans soaked in water was increased by heating them at 100°C for more than 40 min. The 80% methanol extracts of soymilk fermented by *Bifidobacterium adolescentis*, *Bif. breve*, *Bif. longum* and *Lact. lactis* ssp. *lactis* increased the antimutagenic effect compared with that of unfermented soymilk. The extract of fermented soymilk with *Bif. infantis* enhanced the MNNG mutagenesis toward *E. coli*. The antimutagenic effects of the fermented milk with *Bif. bifidum*, *S. salivarius* ssp. *thermophilus* and *L. acidophilus* were lower than that of unfermented soymilk.

Hosoda *et al.* (1992a) investigated the antimutagenic effect of culture milk using 71 strains of lactic acid bacteria belonging to the genera *Lactobacillus*, *Streptococcus*, *Lactococcus* and *Bifidobacteria* on the mutagenicity of MNNG *in vitro* using *Sal. typhimurium* TA 100 as an

Table 5.3 Percentage of inhibition of mutagenesis by extracts of milk and fermented milk.[1] From Nadathur, Gould and Bakalinsky (1994)

Extract	Milk		Streptococcus salivarius ssp. thermophilus[2]		Lactobacillus delbrueckii ssp. bulgaricus[3]		Lactobacillus helveticus[4]		Yoghurt[5]		Milk[6]		Lactobacillus acidophilus[7]	
	X	SEM	X	SEM	X	SEM	X	SEM	X	SEM	X	SEM	X	SEM
(µl)					N-methyl, N'-nitro, N-nitrosoguanidine (%)									
10	10.4^c	0.9	49.3$^{a\,b}$	6.6	29.6^c	2.93	35.3^b	11.2	45.8$^{a\,b}$	14.4	20.4^x	1.3	59.8^y	4.7
20	19.2^b	2.5	77.4^a	7.9	64.3^a	5.3	73.9^a	9.4	80.6^a	5.4	26.5^x	1.8	76.5^y	4.9
50	34.2^b	1.3	89.4^a	5.9	93.4^a	0.8	95.1^a	3.1	93.6^a	2.7	35.4^x	1.5	95.6^y	0.7
					3,2'dimethyl-4-aminobiphenyl (%)									
25	22.9^b	2.3	42^a	4.5	34.3$^{a\,b}$	2.3	33.6$^{a\,b}$	2.7	44^a	6.9	11.1^x	2.8	33.2^y	1.8
50	23.1^c	2.9	58.2^a	3.9	43^b	4	54.5^b	6.4	58.7^a	7.3	19.5^x	2.8	49.5^y	0.6
75	33.9^c	2.8	61.9$^{a\,b}$	5	56.7^b	4.2	64.2$^{a\,b}$	6.1	78.8^a	9.9	24.6^x	1.3	55.1^y	0.8

$^{a\,b\,c;\,x\,y}$ Means in the same row that do not share at least one common superscript are significantly different ($P \leq 0.05$).
Comparisons were (1) between unfermented milk supplemented with 0.5% yeast extract and supplemented milk fermented by *L. acidophilus* BG2F04 (values in last two columns) and (2) between unfermented milk and milk fermented by each of the other organisms (values in first five columns).
[1] Means of three to four experiments.
[2] Strain CH3.
[3] Strain 191R.
[4] Strain CH65.
[5] *S. salivarius* ssp. *thermophilus* CH3 plus *L. delbrueckii* ssp. *bulgaricus* 191R.
[6] Milk supplemented with 0.5% yeast extract (control for *L. acidophilus* BG2F04).
[7] Strain BG2FO4.

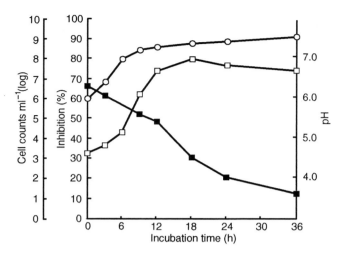

Fig. 5.3 Cell count (○), subsequent pH (■), and inhibition percentage (□) changes of *Lactobacillus acidophilus* LA 106 (LA2). From Hosoda *et al.* (1992a).

indicator bacteria. Each cultured milk sample displayed its characteristic antimutagenic effect on the mutagenicity of MNNG. The milk cultured with *L. acidophilus* LA 106 (LA2) showed the highest inhibition of 77% against the mutagenicity of MNNG among the strains tested. Figure 5.3 shows inhibition percentages for different times of incubation at 36°C. Inhibition percentages rapidly increased with increase in incubation time, reaching a maximum at 18 h of incubation, and gradually decreasing. The inhibitory effects of cultured milk using strains of lactic acid bacteria isolated from milk products on the mutagenicity of 3-amino-1-methyl-5H-pyrido[4,3-b]indole (Trp-P2), a tryptophan pyrolysate for *Sal. typhimurium* TA98, have also been investigated (Hosoda *et al.*, 1992b).

Thyagaraja and Hosono (1993, 1994) evaluated the antimutagenic properties of lactic acid bacteria isolated from 'Idly', a traditional pulse product of southern India. Most of the lactic acid bacteria tested were found to significantly decrease mutagenicities exerted by these mutagens. Time-course studies showed that antimutagenic ability decreased in stationary phase of growth of lactic acid bacteria. There was no correlation between antimutagenicity and enzyme profile quantifying proteolytic, lipolytic and other enzymes of carbohydrate metabolism. In conclusion they found that lactic acid bacteria from 'Idly' were effective in suppressing the mutagenicities of the kinds encountered in foods.

Slimy substance produced with *Brevibacterium linens* exhibited remarkable antimutagenic activity against 3-amino-1,4-dimethy-5H-pyrido[4,3-b]indole (Trp-P1) and (Trp-P2) and oils heated at 160°C for 1 h (Hosono, Yamazaki and Otani, 1989). Lactic acid bacteria such as *Lact. lactis* ssp. *diacetylactis, S. cremoris* and *Leuconostoc paramesenteroides* showed inhibitory effects on the mutagenicities of some volatile nitrosamines (Hosono, Wardojo and Otani, 1990). Hosono, Suzuki and Otani (1991) have found that *S. salivarius* ssp. *thermophilus* 3535 produces a high level of superoxide dismutase which provides a defence mechanism against the toxicity of oxygen.

Contrary to the beneficial effect of lactic acid bacteria in the human intestine, Moore and Moore (1995) have recently reported intestinal floras of populations that have a high risk of colon cancer. They compared the faecal floras of polyp patients, Japanese–Hawaiians, North American Caucasians, rural native Japanese, and rural native Africans (Table 5.4). The polyp patients and Japanese–Hawaiians were considered to be groups at high risk of colon cancer, and the rural native Japanese and rural native Africans were considered to be groups at low risk. The North American Caucasians were found to have a flora composition intermediate between these two groups. Fifteen bacterial taxa from the human faecal flora were significantly associated with high risk of colon cancer, and five were significantly associated with low risk of colon cancer. Total concentrations of *Bacteroides* species and, surprisingly, *Bifidobacterium* species were generally positively associated with increased risk of colon cancer.

A cell wall preparation of *Ent. faecalis* IFO 12965 was very effective for suppression of the induction of mutagenesis in the *E. coli* trp⁻hcr⁻ strain with AF-2 and 4NQO, and was apparently effective with mutagenic faecal extracts from dogs and cats. From these findings, it was suggested that antimutagenic activity of the cell wall fractions could possibly be due to binding mutagens to carbohydrate moieties of the cell wall (Hosono, Yoshimura and Otani, 1987). Tanabe, Otani and Hosono (1991) investigated binding ability of peptidoglycan prepared from the cell wall of *Leu. mesenteroides* ssp. *dextranicum* T-180 to Trp-P1. The results showed that the peptidoglycan preparation had high binding abilities to Trp-P1.

Binding mechanisms are not necessarily the same for the different strains examined. For instance, Morotomi and Mutai (1986) reported that the binding mechanism of Trp-P2 to cell walls of a *Lactobacillus* strain was effected by cation exchange, but Hosono, Yoshimura and Otani (1988) suggested that the binding mechanism of cell walls of *S. faecalis* was not ionic in nature. Tanabe, Suyama and Hosono (1994) investigated the effect of sodium dodecyl sulphate (SDS) on the binding of cells of *Lact. lactis* ssp. *lactis* T-80 isolated from Mongolian kefir with Trp-P1, in order to determine the binding mechanism of the intact cells of this strain with Trp-P1 and observed that the binding of Trp-P1 to

Table 5.4 Species significantly associated with high-risk group (polyp patients and Japanese–Hawaiians). From Moore and Moore (1995)

| Species | Total no. | | P of high | % of isolates in population[b] | | | | |
	Isolates	Sample	vs. low[a]	PP	JH	CAU	NJ	NA
Bacteroides vulgatus	417	67	0.0001	13.12	9.37	8.21	7.52	0.48
Eubacterium rectale 1	75	20	0.0013	2.68	1.56	2.39	0.08	0.1
Ruminococcus torques	27	15	0.0018	0.65	1.33	0.68	0.08	
Streptococcus hansenii	15	11	0.0091	0.65	0.67	0.1	0.08	
Bifidobacterium longum	60	24	0.01	1.57	1.22	2.64	0.37	
Ruminococcus albus	29	13	0.018	1.02	1.11	0.78		
Peptostreptococcus products 1	12	9	0.018	0.46	0.11	0.59		
Bacteroides stercoris	102	32	0.018	3.42	3	1.56	1.53	0.19
Bifidobacterium angulatum	52	14	0.023	2.31	1.89	0.1	0.69	
Eubacterium eligens	70	25	0.024	1.29	3	1.47	0.84	0.29
Eubacterium eligens 2	12	8	0.027	0.37	0.44	0.39		
Ruminococcus gnavus	90	30	0.028	3.23	2.78	0.59	1.53	0.38
Fusobacterium prausnitzii	241	63	0.036	6.01	7	3.03	3.33.74	
Eubacterium cylindroides	10	6	0.037	0.37	0.22	0.39		
Eubacterium rectale 2	166	41	0.042	4.16	3.22	5.87	0.77	2.11
Total no. of isolates				1082	900	1023	1303	1042
Total no. of samples				18	15	17	22	16

[a]Probability that the observed difference between the high- and low-risk groups could occur by chance alone.
[b]PP, Polyp patients; JH, Japanese–Hawaiians; CAU, North American Caucasians; NJ, Rural native Japanese; NA, rural native Africans.

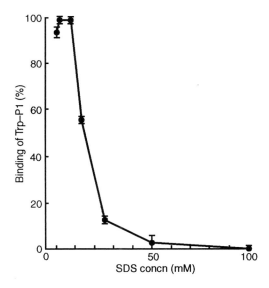

Fig. 5.4 Effect of sodium dodecyl sulphate (SDS) on Trp-P1 to the cells of *Lactobacillus lactis* ssp. *lactis* T-80. Values are means with SE indicated by vertical bars. From Tanabe, Suyama and Hosono (1994).

the cells was significantly inhibited ($P < 0.05$) when SDS was added to the assay system (Figure 5.4). These observations strongly suggest the possibility that the binding of Trp-P1 may be mediated by a cell protein which is denatured by SDS, and involves the formation of hydrophobic bonds. Further, Hosono and Hisamatsu (1995a) examined binding potential of *Ent. faecalis* FK-23 from human gut towards mutagens/ carcinogens like amino acid pyrolysates and aflatoxins, and reported that the binding ability towards aflatoxins B1, B2, G1 and G2 was less compared with the significant binding toward Trp-P1 and Trp-P2, but higher percentage binding toward these aflatoxins was observed when the amount of the cell preparation was increased (Figure 5.5).

Hosono and Tono-Oka (1995b) have recently tested for abilities of lactic acid bacteria from various fermented milk products to bind cholesterol. Of the strains examined the highest binding was observed in *Lact. lactis* ssp. *lactis* biovar. *diacetylactis* R-43. As shown in Figure 5.6, highest binding by this strain was observed immediately after the cells of this strain were incubated with cholesterol. No significant effect of incubation temperatures on the binding of cholesterol to the cell preparation was observed with incubation temperature between 10 and 70°C. Cations such as Mg^{2+} and Ca^{2+} demonstrated a high inhibitory effect on the binding of cholesterol to R-43 strain cells. From this study, it can be concluded that lactic acid bacteria are also effective in binding

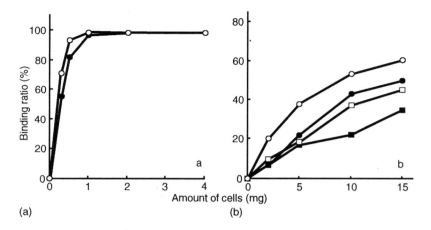

(a)   (b)

Fig. 5.5   Dose–response curves for the binding of amino acid pyrolysates and aflatoxins to cells of *Enterococcus faecalis* FK-23. Trp-P1 (○), Trp-P2 (●) (a), and aflatoxin B_1(○), B_2(●), G_1(□) or G_2(■) (b) was mixed with various amounts of lyophilized cells of *Ent. faecalis* FK-23 in distilled water, and the reaction mixture was incubated at 37°C for 1 h. From Hosono and Hisamatsu (1995a).

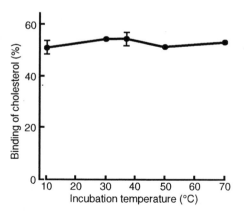

Fig. 5.6   Effect of incubation temperature on the binding of cholesterol to *Lactococcus lactis* ssp. *lactis* biovar. *diacetylactis* R-34. Cholesterol (100 μg) was mixed with 30 mg of the cells in 60% ethanol (1 ml), and the reaction mixture was incubated for 1 h. Error bars denote 95% confidence interval of the means. From Hosono and Tono-Oka (1995b).

cholesterol *in vitro*. This work has an important bearing in the role of therapeutic qualities of fermented milk products, because elevated levels of cholesterol are considered to be a risk factor for coronary heart disease, and also a factor inducing colon cancer in addition to high dietary fat and low fibre (Lichtenstein and Goldin, 1993).

5.3   ANTITUMOUR ACTIVITY OF LACTIC ACID BACTERIA

It is well known that various species of beneficial micro-organisms derived from an animal's natural gut flora have an antitumour activity *in vivo* against experimental tumours (Fernandes *et al.*, 1987; Adachi, 1992). Several researchers revealed that lactic acid bacteria present in milk and milk products, and their cellular components exert an antitumour activity on ascitic and the solid form of a number of transplanted tumours (Friend and Shahani, 1984a; Bounous, Batist and Gold, 1991). The antitumour activity can be broadly categorized as prophylactic effect and therapeutic effect. The prophylactic effect lowers the potential risk of carcinogenesis through elimination and modulation of procarcinogens in the intestinal tract of the host. The therapeutic effect inhibits the growth of tumours or tumour cells, prevents metastasis and induces the complete regression of tumour in the model systems of animals.

Lactic acid bacteria are Gram-positive organisms and some of them inhabit the animal intestine. It has been widely accepted that these bacteria are non-pathogenic and cause few side-effects when injected into animals irrespective of the injection route and they are therefore considered to be suitable for human cancer therapy. Some of the lactic acid bacteria preparations have been used in cancer patients (Bogdanov, 1982; Okawa *et al.*, 1989; Masuno *et al.*, 1991). It is of interest to compare the effect that lactic acid bacteria show to the therapeutic antitumour effect associated with known biological response modifiers (BRMs) such as Bacillus Calmette–Guerin (BCG) (Zbar *et al.*, 1972), *Corynebacterium parvum* (Lichtenstein *et al.*, 1984), *S. pyogenes* (Ebina *et al.*, 1986) and PSK (Ebina and Murata, 1992). Many researchers have studied the biological basis of the antitumour effect of dietary lactic acid bacteria in various animal models for human cancer. The bacterial preparations have been injected either intraperitoneally (i.p.), intravenously (i.v.), subcutaneously (s.c.), intratumourally (i.t.) or orally (p.o.) against the ascitic and solid forms of various types of tumours. During the past decade, the rational understanding of therapeutic role on antitumour activity of dietary lactic acid bacteria in each experimental system has increasingly led to the conclusion that they are acting as BRMs that modulate immune responses *in vivo*. However, the mechanisms operative in the antitumour effect are not the same for different types of tumours. In addition, therapeutic antitumour effect by lactic acid bacteria differs as a function of dosages and times injected, and the route administered.

This chapter focuses on recent therapeutic benefits relating to the antitumour activity of lactic acid bacteria preparations, and details the comparative antitumour activity according to the injection of different routes and schedules in various experimental models. It is hoped that this chapter provides a rationale for further studies to elucidate the mechanisms of therapeutic antitumour action of dietary lactic acid bacteria.

5.3.1   Antitumour activity by intraperitoneal (i.p.) injection of lactic acid bacteria preparations

Several researchers have noted that i.p. injection of lactic acid bacteria exerted an antitumour activity in which the activity was assessed by an increase in the period of survival in ascitic form tumours and by growth inhibition of solid form tumours (Kohwi *et al.*, 1978; Kato *et al.*, 1981; Abe *et al.*, 1982; Shimizu *et al.*, 1987; Kelkar, Shenoy and Kakji, 1988; Kitazawa, Itoh and Yamaguchi, 1991a). The therapeutic antitumour effect of lactic acid bacteria in several experimental models, where the bacteria are injected i.p. after tumour inoculation, is shown in Table 5.5.

Kohwi *et al.* (1978) observed an antitumour activity of *Bif. infantis* of human origin against the ascitic form of syngeneic Meth A fibrosarcoma with an inoculum size of 5×10^4–5×10^5 tumour cells in Balb/c mice. The i.p. injection (10^9 cells) six times daily for 1 week from the day after the tumour cell inoculation suppressed markedly the tumour growth in mice with a small inoculum size (5×10^4 cells) but was less effective on the tumour growth in mice injected with a large inoculum (5×10^5 cells). In addition the mice inoculated with 1×10^5 tumour cells were completely cured of the tumour. The percentage of cured mice decreased as the dosage of the bacteria decreased. The results showed that therapeutic antitumour activity of *Bif. infantis* depended on the dose injected and the number of tumour cells inoculated. In this experiment, killed *Bif. infantis* was as effective as the living one. Therefore, it was stated that the viability or growth of bacteria in the recipient animals could not be an essential factor for its antitumour effect.

Kato *et al.* (1981) observed that the daily i.p. injection (0.2, 2.0, 10.0 mg kg^{-1}) of *L. casei* YIT 9018 for +1 to +5 days from the day after tumour inoculation exhibited marked antitumour activity against the ascitic form of allogeneic S-180 fibrosarcoma in ICR mice and inhibited the growth of methylcholanthrene (MCA)-induced syngeneic MCA K-1 solid tumour in Balb/c mice. However, the capacity of antitumour

Table 5.5 Therapeutic antitumour effect of intraperitoneal injection of lactic acid bacteria

Material subjected	Injection schedule		Tumour model			Antitumour effect[*1]	References
	Doses	Times injected	Tumour type	Inoculum size (× cells) & tumour form	Test subjects		
Lactobacillus bulgaricus	32×10^6 CFU	4	Ehrlich ascites carcinoma	1×10^6 ascites	Swiss mice	ITCP	Friend *et al.*, 1982
L. gasseri	250 µg/mouse	5	Meth A fibrosarcoma	1×10^6 ascites	BALB/c mice	IMST	Shimizu *et al.*, 1987
			Meth A fibrosarcoma	5×10^5 solid	BALB/c mice	ITG	Shimizu *et al.*, 1987
L. casei	0.2–10 mg kg^{-1}	5	Sarcoma180 fibrosarcoma	1×10^6 ascites	ICR mice	IMST	Kato *et al.*, 1981
	1.0–50 mg kg^{-1}	5	L-1210	5×10^4 ascites	BDF1 mice	IMST	Kato *et al.*, 1981
			MCA K-1	1×10^6 solid	BALB/c mice	ITG	Kato *et al.*, 1981
	250 µg mouse^{-1}	5	Meth A fibrosarcoma	1×10^6 ascites	BALB/c mice	IMST	Shimizu *et al.*, 1987
	250 µg mouse^{-1}	5	Meth A fibrosarcoma	5×10^5 solid	BALB/c mice	ITG	Shimizu *et al.*, 1987
L. mesenteroides	4–40 mg kg^{-1}	5	DBDN-induced fibrosarcoma	1×10^5 solid	Swiss mice	IMST	Kelkar *et al.*, 1988
Lactococcus lactis ssp. *cremoris*	10–50 mg kg^{-1}	9	Sarcoma180 fibrosarcoma	1×10^6 ascites solid	ICR mice	ITG	Kitazawa *et al.*, 1991c
Streptococcus salivarius ssp. *thermophilus*	4–40 mg kg^{-1}	5	DBDN-induced fibrosarcoma	1×10^5 solid	Swiss mice	IMST	Kelkar *et al.*, 1988
Lact. lactis ssp. *lactis*	32×10^6 cfu	4	Ehrlich ascites carcinoma	1×10^6 ascites	Swiss mice	ITCP	Friend *et al.*, 1982
	4–40 mg kg^{-1}	5	DBDN-induced fibrosarcoma	1×10^5 solid	Swiss mice	IMST	Kelkar *et al.*, 1988
	4–4 mg kg^{-1}	5	Sarcoma 180 fibrosarcoma / Ehrlich ascites carcinoma / DBDN-induced fibrosarcoma	1×10^6 ascites; 1×10^5 solid	Swiss mice	IST	Kelkar *et al.*, 1988
Bifidobacterium infantis	10^9 cells	6	Meth A fibrosarcoma	5×10^4 ascites	BALB/c mice	ICT,IMST	Kohwi *et al.*, 1978
	2.5 mg kg^{-1}	7	Ehrlich ascites carcinoma	1×10^6 ascites	ICR mice	ICR, IMST	Sekine *et al.*, 1994
S. faecalis	5–50 mg kg^{-1}	7	Ehrlich ascites carcinoma	5×10^4 ascites	DDI mice	ICR, IMST	Abe *et al.*, 1982
			Sarcoma 180 fibrosarcoma	5×10^4 ascites			
	50 mg kg^{-1}	7	Meth A fibrosarcoma	1×10^6 solid	BALB/c mice	IGR	Abe *et al.*, 1982
			MCA fibrosarcoma	5×10^6 solid	C3H mice		
L. fermentum / *L. helveticus*	250 µg mouse^{-1}	5	Meth A fibrosarcoma	1×10^6 ascites	BALB/c mice	IMST	Shimizu *et al.*, 1987
	250 µg mouse^{-1}	5	Meth A fibrosarcoma	5×10^5 solid	BALB/c mice	ITG	Shimizu *et al.*, 1987
L. brevis / *L. celloviosus*	250 µg mouse^{-1}	5	Meth A fibrosarcoma	5×10^5 solid	BALB/c mice	ITG	Shimizu *et al.*, 1987

*1: ITCP, inhibition of tumour cell proliferation; IMST, increase mean survival time; ITG, inhibition of tumour growth; ICR, increase cure rate; IGR, inhibition of growth rate. DBDN, dimethylbenz-dithionaphthene.

activity did not always increase with the dosage. It was also observed that mice given the daily i.p. injection at $10\,\mathrm{mg\,kg^{-1}}$ from -5 to -1 days before tumour inoculation showed the increase of the survival times and rates when allogeneic S-180 fibrosarcoma cells or syngeneic leukaemia L1210 cells were i.p. inoculated. The results showed that *L. casei* 9018 possessed both therapeutic and prophylactic antitumour activity on not only allogeneic but also syngeneic tumour. The activity on syngeneic tumour was less than that of allogeneic tumour. In contrast, the i.p. daily preinjection for 7 days with $250\,\mu\mathrm{g}$ of heat-killed *L. casei* (ATCC 7469) did not exhibit any antitumour activity in either the tumour take or the mean survival time when dimethybenzdithionaphthene (DBDN)-induced fibrosarcoma was inoculated in Swiss mice (Kelkar, Shenoy and Kakji, 1988). However, it remains unclear whether or not the difference of prophylactic effect by the pre-injection of *L. casei* depends on the strain of bacteria or the tumour type inoculated.

In a subsequent study, Shimizu *et al.* (1987) noted that the antitumour activity of four strains of *L. casei* correlated with the capacity to elevate the serum level of colony-stimulating activity (CSA). A dose of $250\,\mu\mathrm{g}$ mouse^{-1} of heat-killed *L. casei* 9018, 0151, 0123 and 0105 was repeatedly injected i.p. on day 3, 6, 9, 12 and 15 after subcutaneous (s.c.) inoculation or on day 1, 3, 5, 7 and 9 after i.p. inoculation of syngeneic Meth A tumour cells in Balb/c mice. For the assay of CSA activity in serum, the bacterial preparations were injected i.p. at a dose of $500\,\mu\mathrm{g}$ mouse^{-1} and serum was obtained 6h after the injection. For *L. casei* 9018, 0151 and 0123, the antitumour activity and CSA-inducing activity showed a good correlation. Similar results were obtained on *L. gasseri* YIT 0075 and 0163, *L. fermentum* YIT 0159, *L. helveticus* YIT 0085 and YIT 0083, *L. brevis* YIT 0076 and *L. cellobiosus* YIT 0079. The increased levels of CSA were also detected in the culture supernatant of peritoneal cells which were obtained 1h after i.p. injection of *L. casei* 9018. Over 95% of the peritoneal cells were macrophages. The CSA-inducing activity in serum was recognized in athymic nu/nu mice and was diminished in the mice pretreated with carrageenan, a selective macrophage blocker. Therefore, it is strongly suggested that CSA plays important roles in therapeutic antitumour activity of *L. casei* 9018 and macrophages serve as potent effecters to tumour cells.

Using an experimental model in the thoracic cavity of mouse suitable for lung cancer, Matsuzaki, Yokokura and Mutai (1988) observed that *L. casei* 9018 had an antitumour activity against Meth A tumour. In the model, Balb/c mice received multiple intrapleural (i.pl.) injections of *L. casei* 9018 at a dose of 100 or $250\,\mu\mathrm{g}$ mouse^{-1} with a different schedule before and after i.pl. inoculation of the tumour cells. The i.pl. post-injection on days +1 and +5 or +1, +3 and +5 of the tumour cell inoculation prolonged the survival of mice and reduced the number of tumour cells in the thoracic cavities. The multiple preinjection with *L.*

casei 9018 on days −5 to −1 or on day −5, −3 and −1 of the tumour cell inoculation also prolonged the survival of mice. In addition, a single injection of *L. casei* 9018 before and after the tumour inoculation caused the antitumour activity and a dose of 100 µg mouse was more effective than 250 µg mouse^{-1} although the activity was less than that of multiple injections. The results confirmed that *L. casei* 9018 has both therapeutic and prophylactic antitumour activity in the thoracic cavity as well as the peritoneal cavity as mentioned above. Thus, it was ascertained that *L. casei* 9018 exerted therapeutic and prophylactic antitumour activity against the ascitic form of various kinds of tumour, S-180 fibrosarcoma, leukaemia L1210 and Meth A fibrosarcoma. When the activity of the multiple injection of *L. casei* 9018 on the survival of Meth A-bearing mice was compared with that of OK432, *Cor. parvum* or BCG, the bacteria were as effective as their BRMs. On the other hand, i. pl. injection of *L. casei* 9018 augmented cytolytic activity of thoracic macrophages and natural killer (NK) cell activity of thoracic exudate cells, and further enhanced phagocytic activity against sheep red blood cells (SRBC) and antigen expression of thoracic macrophages. The fndings suggested that the thoracic macrophages induced by i. pl. injection of *L. casei* 9018 play an important role as one of the main effecter cells for killing tumour cells in the thoracic cavity.

Kelkar, Shenoy and Kaklij (1988) described an antitumour activity of i.p. injection of *S. salivarius* ssp. *thermophilus, Leu. mesenteroides, Lact. cremoris* and *Lact. lactis* against three different types of tumour in Swiss mice. A dose of 4–40 mg kg^{-1} of the heat-killed lactic acid bacteria was daily injected for 5 consecutive days after the tumour inoculation. For solid form of syngeneic DNBA-induced fibrosarcoma, the multiple i.p. injection at a dose of 20 mg kg^{-1} of *S. salivarius* ssp. *thermophilus* prolonged the survival time of the tumour-bearing mice by about 2.0 times but did not result in the complete cure of the mice. Similar results were obtained with i.p. injection of *Leu. mesenteroides, Lact. cremoris* and *Lact. lactis*. Among four strains of the lactic acid bacteria, *S. salivarius* ssp. *thermophilus* appeared to be most effective. In the experiment employing the ascitic form of sarcoma-180 fibrosarcoma and Ehrlich ascites carcinoma, when *S. salivarius* ssp. *thermophilus* was repeatedly injected with the same schedule, a dose of 20 mg kg^{-1} resulted in a 100% cure in sarcoma-180 fibrosarcoma and in only 40% cure in Ehrlich ascites carcinoma. The cured tumours did not recur at the end of 150 days of the experiments. The results showed that the increase of cure rate by the multiple i.p. injection of *S. thermophilus* was dependent on the type and stage of tumour inoculated. However, the mechanism of antitumour effect which was exerted by i.p. injection of four kinds of lactic acid bacteria have not been detailed.

Kitazawa, Itoh and Yamaguchi (1991a) isolated *Lact. lactis* ssp. *cremoris* KVS20 from Scandinavian ropy sour milk 'viili' and defined the antitumour activity against the ascitic and solid form of sarcoma-180 fibrosarcoma. Daily i.p. injection of heat-killed *Lact. cremoris* KVS 20 at a dose of 10 or $50 \, mg \, kg^{-1}$ for 9 successive days from the day after the tumour inoculation inhibited the growth of both ascitic and solid tumour. The treatment was more effective for ascitic-form tumour than for solid-form tumour. The antitumour activity was enhanced as the dosage increased. The dose of $50 \, mg \, kg^{-1}$ prolonged the survival time of mice by about 2.4 times in the ascitic-form tumour and reduced the tumour growth in 48% in solid-form tumour compared with the controls.

In addition to the lactic acid bacteria described above, it was ascertained that *L. delbrueckii* ssp *bulgaricus* (Friend, Fermer and Shahani, 1982) and *S. faecalis* (Abe *et al.*, 1982) also possessed an antitumour activity when repeatedly injected i.p.

As to antitumour mechanisms by i.p. injection of lactic acid bacteria there are several possibilities to be considered:

1. lactic acid bacteria are cytotoxic for tumour cells;
2. non-specific local inflammatory reaction against lactic acid bacteria develops a host-mediated immunological response to the tumour;
3. lactic acid bacteria give rise to the specific immunity to the tumour.

The first possibility is ruled out since several kinds of lactic acid bacteria, *L. casei* 9018 (Kato *et al.*, 1981), *Bif. infantis* (Kohwi *et al.*, 1978), *Lact. lactis* ssp. *cremoris* (Kitazawa, Itoh and Yamaguchi, 1991a) and *S. faecalis* (Abe *et al.*, 1982), have no direct cytotoxic effect on the viability of tumour cells *in vitro*.

With respect to the second possibility, many studies in which i.p. injection of lactic acid bacteria modulates host immune responses in the peritoneal cavity have been described. As shown in Table 5.6, lactic acid bacteria enhance the enzymatic and phagocytic activity of peritoneal macrophages (Hashimoto *et al.*, 1984; Perdigon *et al.*, 1986b) and augment the production of cytokine such as interferon (IFN) (Yamaguchi *et al.*, 1981, 1984; Abe *et al.*, 1982; Kuwabara *et al.*, 1988; Kitazawa *et al.*, 1992) or macrophage activiating factor (MAF) (Kuwabara *et al.*, 1988) from peritoneal exudate cells and the production of oxygen radical (Hashimoto *et al.*, 1984, 1985) or cytostatic factor (Hashimoto *et al.*, 1985, 1987) when injected i.p. It has been well known that cytokine augments the activity of NK cells, activated macrophages and cytotoxic T cells which are involved in the killing of tumour cells. Activated macrophages are able to produce and release oxygen radical or cytostatic factor which play an important role in the tumoricidal actions of macrophages (Hashimoto *et al.*, 1984, 1985). In addition, lactic acid bacteria given i.p. induced cytotoxic macrophages (Hashimoto *et al.*, 1984; Matsuzaki, Yokokura and Mutai, 1988) or NK cells (Matsuzaki, Yokokura and Mutai, 1988; Yamaguchi *et al.*, 1981) which carry out the

Table 5.6 Immunopotentiating effects by i.p. injection of lactic acid bacteria

Parameter of immune responses	Material tested	References
Macrophage function		
CSA production	L. casei 9018,0151,0123	Shimizu et al., 1987
	L. gasseri, L. fermentum	
	L. helveticus, L. celloviosus	
	L. brevis	
Oxygen radical production	L. casei 9018	Hashimoto et al., 1984, 1985
Suppression of PGE$_2$ production	L. casei 9018	Hashimoto et al., 1984
Cytotoxic macrophage induction	L. casei 9018	Kato et al., 1985; Matsuzaki et al., 1988
	Lact. lactis ssp. cremoris	Kitazawa et al., 1991a, b
MAF production	L. casei 9018	Kuwabara et al., 1988
Cytostatic factor production	L. casei 9018	Hashimoto et al., 1985, 1987
Phagocytic activity	L. casei 9018	Kato et al., 1983
	L. acidophilus, S. salivarius ssp. thermophilus	Perdigon et al., 1987
Ia-antigen expression	L. casei 9018	Kuwabara et al., 1988
Enzyme release	L. casei 9018	Kato et al., 1983
	L. acidophilus, S. salivarius ssp. thermophilus	Perdigon et al., 1987
NK cell activity	L. casei 9018	Matsuzaki et al., 1988
	S. faecalis	Yasmaguchi et al., 1981
IFN induction	L. casei 9018	Kuwabara et al., 1988
	S. faecalis	Yamaguchi et al., 1981; Abe et al., 1982
	L. acidophilus	Kitazawa et al., 1992
	L. gasseri	Kitazawa et al., 1994
		Matsumura et al., 1992
Induction of IgM-secreting cells	L. acidophilus, S. salivarius ssp. thermophilus	Perdigon et al., 1987
T-cell dependent antitumour activity	L. casei 9018	Kato et al., 1988

CSA, colony-stimulating activity; MAF, macrophage-activating factor; IFN, interferon, NK, natural killer.

lysis of tumour cells. From these findings, it is supposed that the majority of the antitumour effect by i.p. injection of lactic acid bacteria depends on the macrophage function elicited in the peritoneal cavity.

For the third possibility, it has been ascertained that i.p. injection of *L. casei* 9018 induces Ia-positive macrophage response and a tumour-specific T-cell mediated antitumour immunity in the peritoneal cavity (Kato, Yokokura and Mutai, 1988). When Balb/c mice which were injected i.p. 500 µg mouse $^{-1}$ of *L. casei* 9018 were immunized with i.p. injection of 5×10^6 cells of Meth A fibrosarcoma-treated mitomycin C (MMC-Meth A) and were challenged by i.p. injection of viable Meth A fibrosarcoma (1×10^6 cells) or RL♂ 1 tumour (1×10^6 cells), the growth of Meth A tumour was suppressed and the survival time of the mice was prolonged but the growth of RL♂ 1 tumour was not inhibited. The finding showed that the antitumour immunity induced by i.p. injection of *L. casei* 9018 was tumour specific. The antitumour immunity in the peritoneal cavity was reduced when peritoneal exudate cells (PEC) from the immunized mice were treated with anti-Thy1.2 antibody and complement. Thus, *L. casei* 9018 was able to induce tumour-specific immunity mediated by T cells in the peritoneal cavity of the immunized mice. However, there is no direct evidence that multiple i.p. injection of *L. casei* 9018 after i.p. inoculation of tumour cells gives rise to the induction of tumour-specific immunity in the peritoneal cavity. This is considered to be due to the insufficiency of effecter cells and antigenic stimulation for the development of tumour-specific immunity in the peritoneal cavity.

5.3.2   Antitumour activity by intravenous (i.v.) injection of lactobacillus preparations

Several researchers have attempted to determine whether or not i.v. injection of lactic acid bacteria inhibits solid tumours inoculated s.c. or into the footpad. Kato *et al.* (1981) reported that the growth of allogeneic sarcoma-180 fibrosarcoma in ICR mice was strongly inhibited by multiple daily i.v. injections of *L. casei* 9018 on day +1 to +5. Similar results were obtained in syngeneic MCA K-1 solid tumour in Balb/c mice although the growth inhibition of syngeneic tumour was less than that of the allogeneic tumour. On the other hand, Yasutake *et al.* (1984) observed that a single i.v. injection of *L. casei* 9018 at a dose of 100 µg mouse^{-1} failed to inhibit syngeneic Meth A fibrosarcoma inoculated into the footpad of Balb/c mice. The antitumour activity by multiple i.v. injections of *L. casei* 9018 was reduced in mice which were given carrageenan before the injection of the bacteria but was retained in athymic nu/nu mice. The results showed a possibility that the antitumour effect of i.v. injection of *L. casei* 9018 is macrophage dependent rather than T-cell-dependent.

Asano, Karasawa and Takayama (1986) investigated the possibility of using the antitumour effect produced by i.v. injection of *L. casei* 9018 against MBT-2 murine bladder tumour as an animal model for human bladder cancer. They demonstrated that multiple i.v. injections of heat-killed *L. casei* 9018 at a dose of 0.2 mg mouse^{-1} on the 7th day of tumour inoculation for 10 consecutive days suppressed the tumour growth and delayed the appearance of lung metastasis. However, the mechanisms of antitumour action of the i.v. injection have not been defined. Matsuzaki, Yokokura and Azuma (1985, 1987) examined anti-metastatic activity by i.v. injection of *L. casei* 9018 in C57BL/6 mice which were inoculated with Lewis lung carcinoma (3LL) or a highly metastatic variant of the B16 melanoma cell line (B16-BL6). When a heat-killed preparation of *L. casei* 9018 was repeatedly injected i.v. at a dose of 50, 100 or 250 µg mouse^{-1} on day 7, 10, 13 and 16 after the tumour-cell inoculation, the survival period of mice was prolonged and the pulmonary metastasis was inhibited (Matsuzaki, Yokokura and Azuma, 1985). They also observed that multiple i.v. injections of *L. casei* 9018 with the same schedule prevented the lung metastasis against B16-BL6 inoculated i.v. as a model of an artificial metastasis (Matsuzaki, Yokokura and Azuma, 1987). The antimetastatic activity was dose dependent and the four times injection beginning on the 7th day after the tumour inoculation was most effective and was similar to that of *Cor. parvum*.

However, a single injection of the bacteria on day 1, 7, or 10 did not prevent pulmonary metastasis or prolong the survival of C57BL/6 mice. When tumoricidal activity of the alveolar macrophages against B16-BL6 cells was assessed in C57BL/6 mice injected i.v. with *L. casei* 9018, the activity was augmented and maintained for 3 days. Furthermore, five daily i.v. injections of *L. casei* 9018 from day 7 to day 11 after the inoculation of Meth A tumour cells suppressed the tumour growth and also increased macrophage colony-forming cells (M-CFC) in the femur and spleen (Nanno, Ohwaki and Mutai, 1986). Therefore, it is considered that i.v. injection of *L. casei* 9018 affects macrophage function and the enhanced macrophage function plays an important role in the host defence mechanisms in preventing metastasis.

On the other hand, i.v. injection of *L. casei* 9018 at a dose of 100 µg mouse^{-1} for 3 consecutive days markedly enhanced NK cell activity of spleen cells (Kato, Yokokura and Mutai, 1984). The NK cell activity was maximum on day 3 and declined thereafter when 250 µg mouse^{-1} of the bacteria was injected i.v. In Meth A-bearing mice, the NK cell activity was augmented two-fold by the i.v. injection compared with the controls. From the results, it is strongly suggested that the antitumour activity by i.v. injection of *L. casei* 9018 is closely associated with the ability to increase NK cell activity.

Kimura *et al.* (1980) reported that *Bif. bifidum* survived and proliferated selectively in three kinds of solid tumour when it was injected i.v. into the tumour-bearing mice. The evidence may aid the selective i.v. therapy for cancer.

5.3.3   Antitumour activity by intratumoral (i.t.) injection of lactic acid bacteria preparations

In estimating therapeutic antitumour activity of lactic acid bacteria, it is very important to examine whether or not the bacteria cause the complete tumour regression of solid tumour by i.t. or intralesional (i.l.) injection into solid tumour. Table 5.7 shows an antitumour effect by i.t. or i.l. injection of lactic acid bacteria against various types of solid tumour.

When *Bif. infantis* and *Bif. adolescentis* (10^9 bacterial cells) were repeatedly injected into Meth A fibrosarcoma at +2, +4, +6 and +8 days after the inoculation at 2.5×10^4 tumour cells, many of the tumours underwent complete or partial regression (Kohwi *et al.*, 1978; Kohwi, Hashimoto and Tamura, 1982). The multiple i.l. injections were as effective as OK-432, a kind of BRM which has been clinically used for cancer therapy. However, the antitumour activity of *Bif. infantis* and *Bif. adolescentis* decreased when the first injection of the bacteria was performed at 3 or 7 days after the tumour inoculation or when a large inoculum size of 5×10^5 tumour cells was used. The results showed that tumour regression by multiple i.l. injections of the bacteria was influenced both by the stage of tumour inoculated and the time at which the bacteria were administered. It was further observed that 63 out of 67 mice cured of Meth A fibrosarcoma by multiple i.l. injection of *Bif. infantis* rejected the rechallenged Meth A tumour cells. Therefore, it is supposed that the cured mice are able to acquire the transplantation immunity to Meth A tumour cells.

As to *Bif. infantis*, active components having an antitumour activity *in vivo* have been characterized (Kohwi, Hashimoto and Tamura, 1982; Sekine *et al.*, 1985; Tsuyuki *et al.*, 1991). Sekine *et al.* (1985) isolated three kinds of distinct cell wall preparations, whole peptidoglycan (WPG), WPG-sonicated and cell wall skeleton (CWS), from heat-killed *Bif. infantis*. The antitumour activity of the components was estimated by measuring the growth rate of the tumour in the tumour suppression test (s.c. injection of 100 µg of each component mixed with 10^5 Meth A tumour cells) and in the tumour regression test (multiple injections of 100 µg mouse^{-1} of each component into the tumour-growing site of Meth A fibrosarcoma).

In the tumour suppression test, the three cell wall preparations not only exhibited a high capacity to suppress the tumour growth, but also reduced the tumour incidence. The suppressive activity of WPG was more potent than that of whole cells or CWS and had a high dosage

Table 5.7 Therapeutic antitumour effect of intratumoural or intralesional injection of lactic acid bacteria against solid tumours

Material subjected	Injection schedule		Tumour type (Inoculum size: cells)	Antitumour activity[2]	References
	Doses	Times injected			
Bifidobacterium infantis	10^9 cells mouse^{-1} (viable, killed[1])	6 (on days 2,4,6,8)	Meth A fibrosarcoma (2.5×10^4)	CR, PR	Kohwi et al., 1978, 1982
	100 µg mouse^{-1} (heat killed)	5 (on days 5–9)	Meth A fibrosarcoma (1.0×10^5)	CR, STG	Sekine et al., 1985
Lactobacillus casei (ATCC 7469)	10 mg kg^{-1} (heat killed)	5 (on days 1–5)	DBDN-induced fibrosarcoma (1.0×10^5)	CR, STG	Kelkar et al., 1988
(LC 9018)	100 µg mouse^{-1} (heat killed)	1 (on day 0)	Meth A fibrosarcoma (1.0×10^6)	STG	Yasutake et al., 1984
	2–10 mg kg^{-1} (heat killed)	4 (on days 7, 10, 13, 16)	B16 melanoma (1.0×10^5)	IMST, STG, IM	Matsuzaki et al., 1987
	50–250µg mouse^{-1} (heat killed)	4 (on days 7, 10, 13, 16)	Lewis lung carcinoma (5.0×10^5) Line-10 hepatoma (5.0×10^6)	IMST, STG, IM	Matsuzaki et al., 1985
Streptococcus lactis *Lactococcus lactis* ssp. *cremoris* *S. salivarius* ssp. *thermophilus* *Lactobacillus mesenteroides*	4–40 mg kg^{-1}	5 (on days 1–5)	DBDN-induced fibrosarcoma	CR, STG	Kelkar et al., 1988
L. acidophilus (kz-1293)	10^6 cell mouse^{-1} (viable, formalin killed)	1 (on day 5) 5 (on days 5–9)	Ehrlich ascites carcinoma (1.0×10^5)	IMST, STG, CR	Sakamoto and Konishi, 1988
S. faecalis	1.0 mg mouse^{-1} (viable, heat killed)	1 (on day 3)	Meth A fibrosarcoma (5.0×10^5)	IMST, SGT, CR	

*1: Killed bacteria were obtained by keeping the living bacteria under aerobic conditions for 3 to 7 days.
*2: CR, complete suppression; PR; partial suppression; STG, suppression of tumour growth; IMST, increase mean survival time; IM, inhibition of metastasis.
DBDN, dimethylbenz-dithionaphthene.

dependency in a range of 10–100 μg. The s.c. injection of 100 μg WPG mixed with tumour cells completely regressed the tumour growth and the inhibitory effect of tumour incidence became detectable when the tumour cells were mixed with 20 μg WPG.

In the tumour regression test, five seperate injections of 100 μg mouse^{-1} of WPG or sonicated WPG on days 6–10 after Meth A fibrosarcoma inoculation induced complete regression against the already established tumour in a high proportion of mice. The efficacy was comparable to that of whole cells. Moreover, WPG was clearly more effective than CWS on the tumour-regressive activity. However, a single injection of 500 μg mouse^{-1} of WPG on day 6 after the tumour inoculation had almost no effect against the established tumour. Thus, it was demonstrated that the induction of therapeutic antitumour effect by the multiple i.l. injections of *Bif. infantis* cell walls and their physical structures clearly correlated.

The results strongly suggest that the physical form of cell wall preparations plays an important role in the expression of the tumour-regressive activity. On the other hand, the mice cured of Meth A fibrosarcoma by 5 i.l. injections of WPG showed systemic resistance to reinoculation of the same tumour cells. In this study, it was confirmed that the multiple i.l. injections of WPG gave rise to the generation of tumour immunity. In addition, although they proposed several possibilities on the *in vivo* role of physical form of cell wall preparations of *Bif. infantis*, their differential mechanisms on the potent therapeutic antitumour effect were not detailed.

Recently, the relative antitumour effect of WPG was defned in flora-bearing (FB) and germfree (GF) Balb/c mice using the tumour suppression test as described above (Tsuyuki *et al.*, 1991). When Balb/c mice were inoculated with a mixture of 10^5 Meth A cells and 100 μg WPG, the tumour incidence was 6–10% in FB mice and 50–78% in GF mice in comparison with the tumour incidence of 100% in control mice. The suppression of tumour growth of WPG was obviously far less in GF mice than FB mice. Thus, it was verified that the presence of a gut flora influenced the tumour incidence. These results have great relevance to the clinical application of WPG. Furthermore, it was revealed that the early inflammatory reaction by polymorphonuclear leucocytes (PMN) in the WPG injection site played a role as the major effecter mechanism. When FB (nu/nu), FB (nu/+) and FB (+/+) mice surviving the suppression test were rechallenged s.c. with Meth A tumour cells, FB euthymic mice rejected the rechallenged tumour while none of nu/nu nice tolerated the second challenge. Thus, it was clear that an antitumour immunity by the s.c. injection of WPG mixed tumour cells was developed in euthymic mice but not athymic nu/nu mice, indicating the T cells are involved in the antitumour immunity.

As to *L. casei* 9018, several researchers demonstrated an antitumour effect against various kinds of solid tumour by the i.t. or i.l. injection. Yasutake *et al.* (1984) examined whether or not the simultaneous injection of *L. casei* 9018 and Meth A tumour cells into the different sites result in the suppression of tumour growth. The bacteria inhibited the tumour growth only when injected at a dose of 100 µg mouse^{-1} into the same site inoculated with the tumour cells. The results indicated that the i.l. injection of *L. casei* 9018 into the tumour site is necessary for the induction of antitumour activity against the solid tumour. In the study, it was indicated that the antitumour effect of *L. casei* 9018 was identical with or somewhat superior to those of BCG and *Cor. parvum*.

The mechanism of an antitumour activity of *L. casei* 9018 was analysed in athymic nu/nu mice and with the Winn test by spleen cells or lymph node cells to examine the participation of functional T cells (Yasutake *et al.*, 1984). When a mixture with 100 µg of *L. casei* 9018 and Meth A tumour cells was inoculated into the right hind footpad of nu/nu mice, the tumour growth was suppressed until the 18th day but later the tumour grew significantly in six out of 10 mice. The surviving nu/nu mice which completely suppressed the tumour growth failed to reject the rechallenge Meth A tumour cells. On the other hand, spleen and lymph node cells derived from euthymic mice which were injected with a mixture of *L. casei* 9018 and Meth A tumour cells caused complete suppression of Meth A tumour growth by the Winn test. Furthermore, when the effecter cells were treated with anti-Thy 1.2 antibody and complement, their tumour cytotoxicity was abolished. On the other hand, the specificity of antitumour immunity in mice which were injected with a mixture of *L. casei* 9018 and Meth A tumour cells was ascertained by rechallenging the tumour cells. The mice which completely suppressed primary Meth A tumour completely rejected the rechallenged Meth A tumour cells but not the other type of syngeneic K234 tumour cells when rechallenged on day 5, day 10 or day 30 after the first tumour inoculation. These findings suggest that tumour cytotoxicity induced by *L. casei* 9018 was due to antitumour immunity which is modulated by T cells and the T-cell-mediated antitumour immunity is generated on about the 5th day and continues until at least the 30th day. From the results, it is concluded that the therapeutic antitumour effect by i.l. injection of *L. casei* 9018 is caused by two types of host immune response: non-T-cell-mediated antitumour activity which is induced early after the tumour inoculation and T-cell-mediated antitumour immunity which is required for the complete suppression of tumour growth.

Further investigations showed that multiple i.l. injections of *L. casei* 9018 were effective in the inhibition of tumour growth and metastasis of tumour in the experimental metastasis models of mice or guinea pigs (Matsuzaki, Yokokura and Azuma, 1985, 1987). The antitumour activity against 3LL and B16-BL6 tumours was more extensive when four i.l.

injections were performed on day 7, 10, 13 and 16 after the tumour cell inoculation. The combination treatment of i.l. and i.v. injection of *L. casei* 9018 was most effective for the inhibition of both lymph node and pulmonary metastasis (Matsuzaki, Yokokura and Azuma, 1985, 1987). In addition, multiple i.l. injections of 400 µg or 1.0 mg of the bacteria were markedly effective for the regression of line-10 hepatoma in strain-2 guinea pigs (Matsuzaki, Yokokura and Azuma, 1985). The i.l. injection before surgical excision of the primary tumour inhibited axillary lymph node metastasis and the i.v. injection after surgical excision of the primary tumour inhibited axillary lymph node and lung metastasis. Intrafootpad injection of *L. casei* 9018 enhanced NK cell activity and cytolytic activity of axillary lymph node cells and the i.v. injection augmented alveolar macrophage-mediated cytotoxic activity. From these findings, it was supposed that the action of antimetastatic effect of *L. casei* 9018 was different between the i.l. and i.v. injection. Thus, although it was ascertained that the different types of cells were activated by i.l. and i.v. injection of *L. casei* 9018, the differential antitumour mechanisms of each injection have not been detailed.

It has been reported that multiple i.t. injections of a heat-killed preparation of *S. salivarius* ssp. *thermophilus, Leu. mesenteroides, Lact. cremoris* and *Lact. lactis* gave an antitumour effect. (Kelkar, Shenoy and Kaklij, 1988). In Swiss mice bearing syngeneic DBDN-induced fibrosarcoma, five i.t. injections of all the four different strains at a dose of 10–20 mg kg^{-1} not only inhibited the tumour growth but also completely regressed the tumour. An optimum dose which was recorded for a 100% cure in the fibrosarcoma was the dose of 20 mg kg^{-1} body weight. The antitumour activity with i.t. injection was more effective than that of the five i.p. injections and showed a dose-dependent increase in the number of mice cured. There was no recurrence of the tumours for 150 days. The pretreatment of *S. salivarius* ssp. *thermophilus* prior to the tumour inoculation could not prevent the tumour growth. The complete regression of tumour induced by multiple i.t. injection of *S. salivarius* ssp. *thermophilus* was abrogated when mice were immunosuppressed by sublethal whole-body irradiation of 4 GY or by i.p. injection of 125 mg kg^{-1} of hydrocortisone acetate prior to the tumour inoculation but was not affected when mice were pretreated with carrageenan (Kaklij *et al.*, 1991). In addition, 57out of 60 mice which were completely cured of DBDN-induced fibrosarcoma by i.t. injection of *S. salivarius* ssp. *thermophilus* rejected the rechallenged fibrosarcoma. Spleen cells from the cured mice could effectively transfer the antitumour activity to the tumour-bearing recipients. The antitumour activity was abolished when spleen T cells from the cured mice were depleted with anti-Thy 1.2 antibody and complement. These findings show that T cells, but not macrophages, are involved in the antitumour immunity of multiple i.t. injection of *S. salivarius* ssp. *thermophilus*.

As mentioned above, several kinds of lactic acid bacteria showed indisputable therapeutic, antitumour effects against experimental solid tumours following i.t. or i.l. injection. In particular, *Bif. infantis*, *L. casei* 9018 and *S. thermophilus* induced tumour-specific immunity which was mediated by T-cell function. The antitumour activity was caused by multiple injections and was strongly influenced by times of injection and the timing of first injection. Sakamoto and Konishi (1988) observed that a single i.t. injection of 11 bacterial strains isolated from intestinal microflora suppressed the growth and appearance of solid Ehrlich ascites tumour when injected on day 5 after the tumour cell inoculation. The activity of a single injection was less than that of five daily injections starting on day 5. By contrast, a single i.l. injection of *L. casei* 9018 on day 1 or 3 was not effective against B16-BL6 tumour (Matsuzaki, Yokokura and Azuma, 1987). A single i.l. injection of 500 µg WPG also had no effect on the established Meth A tumour (Reddy *et al.*, 1983). To date, it has not been determined if a single i.t. injection of lactic acid bacteria exerts an antitumour effect and, if so, whether the antitumour activity depends on the strain of bacteria or the susceptibility of the tumour types.

We have attempted to demonstrate antitumour activity by a single i.t. injection of TH69, a lyophilized preparation of *S. faecalis* TH001. Balb/c mice which have a palpable Meth A fibrosarcoma on day 3 after the tumour cell inoculation into the right flank received a single i.t. injection of TH69 (1 mg mouse^{-1}), BCG (1 mg mouse^{-1}) or mitomycin C (MMC; 0.1 mg mouse^{-1}). The complete regression of Meth A tumour was observed in 9 out of 10 mice by TH69 injection and in 6 out of 6 mice by MMC injection. However, neither significant growth inhibition nor regression of the tumour was obtained in mice injected with BCG (Table 5.8). A single i.t. injection of TH69 was much more effective and showed a dose-dependent increase in the number of mice cured. When a single i.t. injection of TH69 was performed on day 3, 5 or 7 after the tumour inoculation, the injection on day 3 was most effective in which 6 out of 8 mice completely regressed the tumour (TH69-Meth A CR mice), whereas the number of TH69-Meth A CR mice was decreased as the interval of TH69 injection from the time of tumour cell inoculation was prolonged (Figure 5.7). The results indicate that the dosage and timing of TH69 administration are very important in the induction of complete tumour regression.

We further investigated whether or not a single i.t. injection of TH69 produces tumour immunity in the process of inducing the tumour regression. TH69-Meth A CR mice received the intradermal rechallenge with 1×10^6 cells of Meth A fibrosarcoma or RL♂ 1 leukaemia in the flank opposite the primary tumour. Six out of 9 mice completely rejected the rechallenged Meth A tumour cells (TH69-Meth A CR mice). The remaining three mice strongly inhibited the growth of the tumour

Table 5.8 An antitumour activity against Meth A fibrosarcoma by a single TH69 intratumoural injection

Experiment	Injection	Dose (mg)	No. of tumour-free mice/ no. of mice tested	Tumour weight (g $\pm$ SD)
1	PBS	–	0/6	9.0 $\pm$ 2.85
	BCG	0.2	0/6	8.1 $\pm$ 5.20
	BCG	1	0/6	6.1 $\pm$ 4.05
	MMC	0.02	0/6	1.2 $\pm$ 1.15**
	MMC	0.05	2/6	1.8 $\pm$ 1.82**
	MMC	0.1	6/6	
	TH69	0.1	1/6	3.8 $\pm$ 2.44**
2	PBS	–	0/10	11.7 $\pm$ 3.32
	TH69	0.01	0/10	9.1 $\pm$ 3.64*
	TH69	0.1	1/10	4.4 $\pm$ 2.83**
	TH69	1	9/10	1.7 $\pm$ 0.00**

Significant difference from PBS-injected controls. *:$P < 0.05$, **$P < 0.01$ PBS, phosphate buffered saline; BCG, Bacillus Calmette-Guerin; MMC, mitomycin C; TH69, lyophilized preparation of *S. faecalis* TH001.

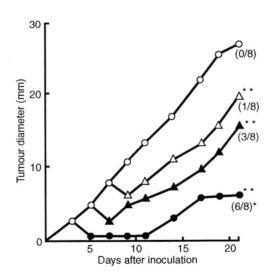

Fig. 5.7 Antitumour activity by a single i.t. injection of TH69. Mice received a single TH69 injection at various times (arrows) after the inoculation of Meth A tumour cells. ○, control; ●, on day 3 injection; ▲, on day 5 injection; △, on day 7 injection. The number in parentheses indicates no. of completely regressed mice/no. of mice tested. Significant difference from control for wet tumour weight on day 21. **: $P < 0.01$, *: $P < 0.05$.

rechallenged. On the other hand, the rechallenged Meth A tumour cells grew in 6 out of 6 mice which completely regressed Meth A tumour by a single injection of MMC (MMC-Meth A CR mice) (Table 5.9). Four out of 6 TH69-Meth A CR mice rejected the Meth A tumour cells challenge. By contrast, when TH69-Meth A CR mice were challenged with RL♂1 tumour cells, the tumour grew continuously in 6 out of 6 mice as it did in control mice (Table 5.10, Exp. 1). In addition, when mice which rejected the rechallenge with Meth A tumour cells were rechallenged again with 1×10^6 Meth A tumour cells, all the mice showed a more definite specific rejection of the second tumour rechallenge. However, the mice could not reject the challenged RL♂1 tumour cells although the growth of RL♂1 tumour was significantly inhibited (Table 5.10, Exp. 2). Thus, it was demonstrated that a clear-cut tumour resistance had occurred as a result of a single i.t. injection of TH69, indicating that tumour-specific immunity is generated concurrent with the tumour regression.

The regression and growth inhibition of Meth A tumour by a single i.t. injection of TH69 was not influenced by pretreatment with carrageenan or trypan blue. However, the antitumour activity of TH69 injection was not observed in athymic nu/nu mice. The findings indicate that T cells, but not macrophages, are required for establishing the tumour regression and tumour-specific immunity in mice which received a single i.t. injection of TH69. Therefore, it is possible to conclude that a single i.t. injection of lactic acid bacteria caused the growth inhibition and regression against established solid tumour. From our study, it is concluded that a successful intratumoural immunotherapy with the induction of tumour-specific immunity by lactic acid bacteria can be applied clinically.

Table 5.9 Development of tumour immunity in mice completely regressed Meth A fibrosarcoma by TH69 intratumoural injection

Injection (dose)	Primary tumour		Rechallenged tumour	
	No. of tumour-free mice /no. of mice tested	Tumour weight (g ± >SD)	No. of tumour-free mice /no. of mice tested[1]	Tumour weight (g ± SD)
PBS	0/6	8.2 ± 2.05	0/6	7.8 ± 3.65
TH69 (1.0 mg)	9/10	2.9 ± 0.00**	6/9**	1.2 ± 0.14**
MMC (0.1 mg)	6/6	–	0/6	2.3 ± 1.82**
BCG (1.0 mg)	0/6	6.6 ± 3.05	–	–

Significant difference from PBS-injected controls. *P < 0.05, **P < 0.01.
[1]No. of mice who completely regressed primary Meth A tumour/no. of mice tested
PBS, phosphate buffered saline; TH69, lyophilized preparation of *S. faecalis* TH001; MMC, mitomycin C; BCG, Bacillus Calmette-Guerin.

Table 5.10 Development of tumour specific immunity in TH69-Meth A CR mice

Experiment	Mice tested	Rechallenged tumour	No. of tumour-free mice /no. of mice tested	Tumour weight (g ± SD)
1	Normal mice	Meth A	0/8	9.2 ± 3.26
	Normal mice	RL♂1	0/8	7.8 ± 2.18
	TH69-Meth A CR-mice[1]	Meth A	4/6[*]	1.8 ± 1.14[**]
	TH69-Meth A CR-mice[1]	RL♂1	0/6	6.8 ± 1.92
2	Normal mice	Meth A	0/8	10.2 ± 3.66
	Normal mice	RL♂1	0/8	
	TH69-Meth A CR-mice[2]	Meth A	6/6[**]	–
	CR-mice[2]	RL♂	0/6	4.4 ± 1.18[*]

Mice who rejected once[1] or twice[2] the challenge with Meth A fibrosarcoma were rechallenged again with Meth A of RL♂1 tumour cells.
Significant difference from PBS-injected controls. [*]P < 0.05, [**]P < 0.01.
TH69, lyophilized preparation of *S. faecalis* TH001

Recently, Ebina *et al.* have devised the double grafted tumour system as a new experimental tumour model on the evaluation of antitumour activity by i.t. injection of BRMs or chemotherapeutic drugs (Ebina *et al.*, 1986; Ebina and Murata, 1992). The model is particularly suited for analysing the generative mechanism of tumour immunity or the effecter mechanism against solid tumour. In the further study of the modes of antitumour action of lactic acid bacteria resembling BRMs, the double grafted tumour system will prove to be a very useful tool.

5.3.4   Antitumour activity by oral (p.o.) administration of lactic acid bacteria preparations

The establishment that oral administration of dairy lactic acid bacteria or fermented milk exerts a therapeutic antitumour effect would provide a very attractive form of therapy in humans. Feeding fermented milk, colostrum or milk cultured with *L. acidophilus* and/or *L. delbrueckii* ssp. *bulgaricus*, or *L. delbrueckii* ssp. *bulgaricus* plus *S. salivarius* ssp. *thermophilus* for 7 days after i.p. inoculation of Ehrlich ascitic tumour cells inhibited the growth rate of tumour cell proliferation (Shahani, Friend and Bailey, 1983). Reddy, Shahani and Banerjee (1973), Reddy *et al.* (1983) and Friend, Fermer and Shahani (1982) observed that feeding of yoghurt or yoghurt components for 7 consecutive days after i.p. inoculation of Ehrlich ascitic tumour cells produced significant antitumour activity. When yoghurt was fed *ad lib* for 7 days before the tumour inoculation until the mice were sacrificed on day 8, prefeeding did not offer an additional antitumour effect. Furthermore, feeding yoghurt from 4 day after the tumour inoculation had no significant antitumour activity. The findings showed that feeding of yoghurt was most effective in the early stages of tumour growth in the peritoneal cavity. The per cent survival rate of tumour-bearing mice did not increase when yoghurt feeding was continued until death occurred. Thus, although yoghurt feeding effectively inhibits initial tumour growth, it does not appear to retard long-term growth of tumour cells.

The basis of the antitumour activity of yoghurt was investigated. The components of yoghurt were tested for inhibition of the growth of Ehrlich ascitic tumour cells in the peritoneal cavity. Ayebo, Shahani and Dam (1981) observed that feeding of the yoghurt dialysate resulted in 32.9% inhibition of the tumour cell count, while feeding of the yoghurt retentate did not reduce the tumour proliferation. They pointed out that the antitumour activity may be due to a component(s) with a molecular weight ⩽ 14 000. On the other hand, the supernatant fluid fraction of yoghurt had no inhibitory effect and the solid fraction of yoghurt inhibited the tumour growth by 22–34% when they were fed for 7 days following i.p. tumour inoculation (Reddy *et al.*, 1983). Feeding of a 3 × concentrate of the solid fraction did not significantly increase the antitumour activity. From these findings, it is most likely that the antitumour activity of yoghurt is due to component(s) other than lactic acid which are produced during yoghurt manufacture and/or storage. However, it is still unclear which component(s) of yoghurt exerts the antitumour activity. Since yoghurt is prepared from a stock culture of *L. delbrueckii* ssp. *bulgaricus* and *S. salivarius* ssp. *thermophilus*, it is expected that oral administration of the bacterial constituent may have a therapeutic antitumour effect. As mentioned above, feeding of yoghurt or yogurt component(s) had an antitumour effect against Ehrlich ascitic tumour in the peritoneal cavity; however, it is not known whether or not the feeding is effective against syngeneic solid tumour.

Bogdanov (1982) found oral administration of extracts prepared from the specific strain of *L. delbrueckii* ssp. *bulgaricus* strain 'LB51' possessed an antitumour effect *in vivo* on transplanted tumours in mice. Recently, Asano, Karasawa and Takayama (1986) observed that a patient with unresectable carcinoma of the bladder who enjoyed drinking Yakult, a lactobacilli-fermented beverage, survived 4.5 years without any evidence of metastasis. Following on from this observation, they investigated the antitumour effect of p.o. administration of viable *L. casei* 9018 against a one-cell line of experimental murine bladder tumour, MDT-2, which is induced in the urinary bladder of female C3H/He mice by

N-[4-(5-nitro-2-furyl)-2 thiazonyl] formamide (FANFT) and easily grows and metastasizes in the lungs when inoculated into the hind limbs of syngeneic mice. Multiple p.o. administration of viable *L. casei* 9018 containing 1.6×10^8 cells per millilitre on the 7th day of tumour inoculation for 10 consecutive days significantly inhibited the tumour growth. The inhibitory effect was also observed when the tumour-bearing mice were fed *ad libitum* the viable bacteria mixed with drinking water from the day of tumour inoculation until the day of sacrifice. Furthermore, p.o. administration of *L. casei* 9018 lowered the incidence of pulmonary metastasis. The histological findings revealed that granular degeneration, cytolysis and necrosis took place in the tumours of mice treated with the bacteria. Thus, it was clearly demonstrated that p.o. administration of viable *L. casei* 9018 exerted a therapeutic antitumour effect against solid tumour. However, the mechanism of action of *L. casei* 9018 given orally was not defined in the experiment.

In order to understand the mechanisms of antitumour action of dairy lactic acid bacteria, it is important to clarify whether or not the p.o. administration modifies the host immune system. Perdigon *et al.* (1986a) reported that *L. casei* and *L. delbrueckii* ssp. *bulgaricus* given p.o. with a dose of $50 \, \mu g \, day^{-1}$ for 8 consecutive days induced the release of lysosomal and non-lysosomal enzymes from peritoneal macrophages of mice. The increased enzymatic activity of macrophages after oral administration began on the 5th day for *L. delbrueckii* ssp. *bulgaricus* and on the 2nd day for *L. casei*. The enzyme activity of peritoneal macrophages by p.o. injection of *L. casei* was higher with viable cells than non-viable cells, while no difference was observed between viable and non-viable cells of *L. delbrueckii* ssp. *bulgaricus* (Perdigon *et al.*, 1986b). The phagocytic activity of *Sal. typhi* by peritoneal macrophages was enhanced only when *L. casei* was orally administered, but did not differ between viable and non-viable cells (Perdigon *et al.*, 1986b). Furthermore, Perdigon *et al.* (1987) observed that *L. casei* and *L. delbrueckii* ssp. *bulgaricus* accelerated the phagocytic function of the reticuloendothelial system in mice given p.o. from day 2 onward. They also noted that p.o. adminstration of *S. salivarius* ssp. *thermophilus*, *L. acidophilus*, *L. casei* and the mixture of *L. casei* and *L. acidophilus* enhanced both enzymatic and phagocytic activities of peritoneal macrophages and activated lymphocytic function as determined by the number of spleen IgM-plaque-forming cells against SRBC (Perdigon *et al.*, 1987). The evidence indicates that lactic acid bacteria given p.o. are capable of activating macrophages. Furthermore, it was reported that multiple p.o. administration at a dose of $150 \, mg \, kg^{-1}$ of DEODAN, a lysozyme lysate from *L. delbrueckii* ssp. *bulgaricus* 'LB51' for oral administration, caused an increase of the spreading ability and phagocytic activity of peritoneal macrophages (Popova *et al.*, 1993). The level of interleukin 1 (IL-1) in the culture supernatant of peritoneal macrophages from mice treated p.o. with DEODAN was slightly increased. Therefore, it is strongly suggested that one of the antitumour mechanisms with p.o. administration of lactic acid bacteria could be non-specific cytotoxicity against neoplastic target cells which was mediated by the activation of macrophage function.

As described above, it was shown that heat-killed *L. casei* 9018 had remarkable antitumour activity when injected i.p., i.v. and i.l. or i.t., but not when orally administered. Recently, Kato, Endo and Yokokura (1994) investigated whether or not p.o. administration of viable *L. casei* 9018 possessed the ability to induce antitumour activity against solid tumours and to modify the host immune system. Balb/c mice were immunized by surgical resection of a primary colon 26 tumour which had been implanted in the abdomen with 5×10^5 cells 5 days previously. Three days later, 1×10^5 cells of secondary colon 26 tumour were injected into the site opposite the primary tumour and Biolactis powder (BLP), a preparation of viable *L. casei* 9018, containing 2.3×10^{11} cells g^{-1}, was repeatedly administered orally for 7 consecutive days at a dose of 50, 100 or 200 mg kg^{-1}day^{-1}. The growth of secondary tumour was markedly suppressed by p.o. administration with BLP at a dose of 100 or 200 mg kg^{-1}day^{-1}. When BLP was killed by heating for 60 min at 70°C, the p.o. administration at the same dose did not exhibit any suppression against secondary tumour. The suppression was a primary tumour-specific response, in which p.o. administration of BLP inhibited the growth of the secondary colon 26 tumour but not the growth of the secondary Meth A tumour. On the other hand, in non-immunized mice, the growth of colon 26 tumour was not affected by p.o. administration of BLP. The findings indicated that the effectiveness of the orally administered BLP was dependent upon antitumour immune responses rather than non-specific inflammatory responses, since secondary tumour rejection was restricted by the resected primary tumour. In the experiment, the responses of splenocytes to T-cell mitogen, IL-1 and IL-2 in the tumour-bearing mice were significantly lower than those in normal mice. The suppressive responses were abolished by p.o. administration of BLP, indicating that oral BLP is able to potentiate systemic immune responses that modify T-cell function in tumour-bearing mice. The study first demonstrated that p.o. administration of lactic acid bacteria enhances T-cell function and induces tumour-specific immunity.

From the studies reviewed above it can be concluded that viable bacteria are more effective than non-viable ones in producing the antitumour effects. It has been accepted that some indigenous enteric organisms can penetrate the barrier and translocate to the internal organs. Serum antibody to viable *Bif. breve* was detected in mice fed the bacteria for 33 days and the antibody production was regulated in Peyer's patches which are groups of subepithelial, lymphoid follicles located throughout the small intestine (Yasui, Mike and Ohwaki, 1989). Viable *Bif. longum* could be easily translocated from the gut epithelial layer to the mesenteric lymph nodes, liver and kidneys in germ-free and T-cell deficient hosts (Yamazaki *et al.*, 1985). Recently, Takahashi *et al.* (1991) histocytochemically demonstrated that viable *L. casei* 9018 were phagocytosed into M cells or macrophages in Peyer's patches in the rabbit and rat ileum. M cells are morphologically characterized by microfolds at the luminal surface (Owen and Jones, 1974) and play a key role in initial intestinal local immune responses by causing trans-epithelial migration of bacterial antigens from the lumen into the lymphoid tissues (Owen *et al.*, 1986). Therefore, it is strongly suggested that viable lactic acid bacteria given p.o. may attach to the intestinal wall and translocate through the intestinal tract to potentiate the host immune system. The difference on a therapeutic antitumour effect of oral administration between viable and non-viable lactic acid bacteria might be explained by the ability of viable bacteria to translocate through the intestinal wall.

5.4   REFERENCES

Abe, N., Yamaguchi, T., Hoshino, F. *et al.*. (1982) Antitumor and interferon-inducing activities of TH69, whole bacterial preparation of *Streptococcus faecalis*, in mice. *Gann*, **73**, 811–18.

Adachi, S. (1992) Lactic acid bacteria and the control of tumours in the lactic acid bacteria, in *The Lactic Acid Bacteria in Health and Disease.*, Vol. 1 (ed. B. J. B. Wood), Elsevier Applied Science, London, pp. 233–61.

Ahmed, F. E. (1992) Effect of diet on cancer and its progression in humans. *Environ. Carcinog. Ecotoxicol. Rev.*, **C10**, 141–80.

Asano, M., Karasawa, E. and Takayama, T. (1986) Antitumour activity of *Lactobacillus casei* (LC9018) against experimental mouse bladder tumour (MBT-2). *J. Urology*, **136**, 719–21.

Ayebo, A. D., Shahani, K. M. and Dam, R. (1981) Antitumour component(s) of yogurt: fractionation. *J. Dairy Sci.*, **64**, 2318–23.

Ayebo, A. D., Shahani, K. M., Dam, R. and Friend, B. A. (1982) Ion exchange separation of the antitumour component(s) of yogurt dialysate. *J. Dairy Sci.*, **65**, 2388–90.

Ballongue, J. (1993) Bifidobacteria and probiotic action, in *Lactic Acid Bacteria* (ed. S. Salminen and A. von Wright), Marcel Dekker, New York, pp. 357–428.

Bodana, A. R. and Rao, D. R. (1990) Antimutagenic activity of milk fermented by *Streptococcus thermophilus* and *Lactobacillus bulgaricus*. *J. Dairy Sci.*, **73**, 3379–84.

Bogdanov, I. (1982) Observation on the therapeutic effect of the anti-cancer preparation from *Lactobacillus bulgaricus* 'LB-51' tested on 100 oncologic patients (ed. I. Bogdanov), Sofia Press, Sofia, pp. 127.

Bounous, G., Batist, G. and Gold, P. (1991) Whey proteins in cancer prevention. *Cancer Lett.*, **57**, 91–4.

Buck, L. M. and Gilliland, S. E. (1994) Comparisons of freshly isolated strains of

Lactobacillus acidophillus of human intestinal origin for ability to assimilate cholesterol during growth. *J. Dairy Sci.*, **77**, 2925–33.

Deeth, H.C. (1984) Yogurt and cultured products. *Aust. J. Dairy Tech.*, **39**, 111–13.

Doll, R. and Peto, R. (1981) The causes of cancer: quantitative estimates of avoidable risks of cancer in the United States today. *J. Natl Cancer Inst.*, **66**, 1191–308.

Ebina, T. and Murata, K. (1992) Antitumour effect of PSK at a distant site: tumour-specific immunity and combination with other chemotherapeutic agents. *Jpn. J. Cancer Res.*, **83**, 775–82.

Ebina, T., Kohya, H., Yamaguchi, T. and Ishida, N. (1986) Antimetastatic effect of biological response modifiers in the 'double grafted tumour system'. *Jpn J. Cancer Res. (Gann)*, **77**, 1034–42.

Fernandes, C. E., Shahani, K. M. and Amer, M. A. (1987) Therapeutic role of dietary lactobacilli and lactobacilli fermented dairy products. *FEMS Microbiol. Rev.*, **46**, 343–56.

Friend, B. A. and Shahani, K. M. (1984) Antitumour properties of lactobacilli and dairy products fermented by lactobacilli. *J. Food Pro.*, **47**, 717–23.

Friend, B. A., Fermer, R. E. and Shahani., K. M. (1982) Effect of feeding and intraperitonial implantation of yogurt culture cells on Ehrlich ascites tumour. *Milchwissenschaft*, **37**, 708–10.

Gibson, G. R. and Wang, Xin. (1994) Regulatory effects of bifidobacteria on the growth of other colonic bacteria. *J. Appl. Bact.*, **77**, 412–20.

Gibson, G. R., Beatty, E. R., Wang, Xin and Cummings, J. H. (1995) Selective stimulation of *Bifidobacteria* in the human colon by oligofructose and inulin. *Gastroenterology*, **108**, 975–82.

Gilliland, S. E. (1989) Acidophilus milk: a review of potential health benefits to consumer. *J. Dairy Sci.*, **72**, 2483–94.

Gilliland, S. E. (1990) Health and nutritional benefits from lactic acid bacteria. *FEMS Microbiol. Rev.*, **87**, 175–88.

Gilliland, S. E. (1991) Properties of yoghurt, in *Therapeutic Properties of Fermented Milks* (ed. R. K. Robinson), Elsevier Applied Science, London, pp. 65–80.

Gilliland, S.E. and Speck, M. L. (1977) Antagonistic action of *Lactobacillus acidophilus* towards intestinal and food borne pathogens in associative culture. *J. Food Prot.*, **40**, 820–23.

Gilliland, S. E. and Walker, D. K. (1990) Factors to consider when selecting a culture of *Lactobacillus acidophilus* as a dietary adjunct to produce a hypocholesterolemic effect in humans. *J. Dairy Sci.*, **73**, 905–11.

Gilliland, S.E., Nelson, C. R. and Maxwell, C. (1985) Assimilation of cholesterol by *Lactobacillus acidophilus*. *Appl. Environ. Microbiol.*, **49**, 377–81.

Gurr, M.I. (1987) Nutritional aspects of fermented milk products. *FEMS Microbiol. Rev.*, **64**, 337–42.

Hamdan, I. T. and Mikolajicik, E. M. (1974) Acidolin: an antibiotic produced by *Lactobacillus acidophilus*. *J. Antibiotics*, **27**, 631–6.

Harrison, V. C. and Peat, G. (1975) Serum cholesterol levels in rats fed milk fermented by *Lactobacillus acidophilus*. *J. Food Sci.*, **47**, 2078–9.

Hashimoto, S., Nomoto, K., Matsuzaki, T. *et al.* (1984) Oxygen radical production by peritoneal macrophages and kupffer cells elicited with *Lactobacillus casei*. *Infect. Immun.*, **44**, 61–7.

Hashimoto, S., Seyama, Y., Yokokura, T. and Mutai, M. (1985) Cytotoxic factor production in kupffer cells elicited with *Lactobacillus* and *Corynebacterium parvum*. *Cancer Immunol. Immunother.*, **20**, 117–21.

Hashimoto, S., Nomoto, K., Nagaoka, M. and Yokokura, T. (1987) *In vitro* and *in vivo* release of cytostatic factors from *Lactobacillus casei* elicited peritoneal macrophages after stimulation with tumour cells and immunostimulants. *Cancer Immunol. Immunother.*, **24**, 1–7.

Hepner, G., Fried, R., Jeor, S. St. *et al.* (1979) Hypocholesterolemic effect of yogurt and milk. *Am. J. Clin. Nutr.*, **32**, 19–24.

Hosoda, M., Hashimoto, H., Morita, H. *et al.* (1992a) Antimutagenicity of milk cultured with lactic acid bacteria against N-methyl-N'-nitrosoguanidine. *J. Dairy Sci.*, **75**, 976–81.

Hosoda, M., Hashimoto, H., Morita, H., Chiba, M. and Hosono, A. (1992b) Studies on antimutagenic effect of milk cultured with lactic acid bacteria on the Trp-P2-induced mutagenicity to TA 98 strain of *Salmonella typhimurium*. *J. Dairy Res.*, **59**, 543–9.

Hosono, A. (1988) The role of lactic acid bacteria as a scavenger of N-nitroso compounds in the intestinal tract. *Bull. Jpn Dairy Tech. Assoc.*, **38**, 1–17.

Hosono, A. (1992) Diet and mutagenic toxicity, in *Functions of Fermented Milk* (ed. Y. Nakazawa and A. Hosono), Elsevier Applied Science, London, pp. 233–61.

Hosono, A. and Hisamatsu, S. (1995a) Binding of amino acid pyrolysates and aflatoxins to autoclaved cells of *Enterococcus faecalis* FK-23. *Biosci. Biotech. Biochem.*, **59**, 940–2.

Hosono, A. and Tono-Oka, T. (1995b) Binding of cholesterol with lactic acid bacterial cells. *Milchwissenschaft*, **50**, 556–60.

Hosono, A., Kashina, T. and Kada, T. (1986) Antimutagenic properties of lactic acid-culture milk on chemical and fecal mutagens. *J. Dairy Sci.*, **69**, 2237–42.

Hosono, A., Sagae, S. and Tokita, F. (1986) Desmutagenic effect of cultured milk on chemically induced mutagenesis in *Escherichia coli* B/r WP2 trp⁻hcr⁻. *Milchwissenschaft*, **41**, 142–5.

Hosono, A., Yoshimura, S. and Otani. H. (1987) Antimutagenic activity of cellular component of *Streptococcus faecalis* IFO 12965. *Netherlands Milk and Dairy J.*, **41**, 239–45.

Hosono, A. Yoshimura, S. and Otani, H. (1988) Desmutagenic property of cell walls of *Streptococcus faecalis* on the mutagenicities induced by amino acid pyrolysates. *Milchwissenschaft*, **43**, 169–70.

Hosono, A., Shashikanth, K. N. and Otani, H. (1989) Antimutagenic activity of whole casein on the pepper induced mutagenicity of streptomycin-dependent strain SD 510 of *Salmonella typhimurium* TA 98. *J. Dairy Res.*, **55**, 435–42.

Hosono, A., Yamazaki, H. and Otani, H. (1989) Antimutagenicity of slimy substance separated from the culture of *Brevibacterium lines*. *Jpn J. Zootech. Sci.*, **60**, 679–85.

Hosono, A., Wardojo, R. and Otani, H. (1990) Inhibitory effects of lactic acid bacteria from fermented milk on the mutagenicities of volatile nitrosamines. *Agr. Biol. Chem.*, **54**, 1639–43.

Hosono, A., Suzuki, M. and Otani, H. (1991) Superoxide dismutase activity in the crude cell extract from *Streptococcus salivarius* subsp. *thermophilus* 3535. *Anim. Sci. Tech.*, **62**, 39–44.

Kada, T., Inoue, T. and Namiki, M. (1981) Environmental desmutagens and antimutagens, in *Environmental Mutagenesis, Carcinogenesis and Plant Biology*, Vol. 1 (ed. E. J. Klekowski), Praeger, New York, pp. 133–51.

Kaklij, G. S., Kelkar, S.M. Shenoy, M. A. and Sainis, K. B. (1991) Antitumour activity of *Streptococcus thermophilus* against fibrosarcoma: role of T-cells. *Cancer Lett.*, **56**, 37–43.

Kanatani, K., Oshimura, M. and Sano, K. (1995) Isolation and characterization of acidocin A and cloning of the bacteriocin gene from *Lactobacillus acidophilus*. *Appl. Environ. Microbiol.*, **61**, 1061–7.

Kato, I., Kobayashi, S., Yokokura, T. and Mutai, M. (1981) Antitumour activity of *Lactobacillus casei* in mice. *Gann*, **72**, 517–23.

Kato, I., Yokokura, T. and Mutai, M. (1983) Macrophage activation by *Lactobacillus casei* in mice. *Microbiol. Immunol.*, **27**, 611–18.

Kato, I. Yokokura, T. and Mutai, M. (1984) Augmentation of mouse natural killer cell activity by *Lactobacillus casei* and its surface antigens. *Microbiol. Immunol.*, **27**, 209–17.

Kato, I., Yokokura, T. and Mutai, T. (1985) Induction of tumoricidal peritoneal exudate cells by administration of *Lactobacillus casei*. *Int. J. Immunopharmac.*, **7**, 103–9.

Kato, I., Yokokura. T. and Mutai, M. (1988) Correlation between increase in Ia-bearing macrophages and induction of T cell-dependent antitumour activity by *Lactobacillus casei* in mice. *Cancer Immunol. Immunother.*, **26**, 215–21.

Kato, I., Endo, K. and Yokokura, T. (1994) Effects of oral administration of *Lactobacillus casei* on antitumour responses induced by tumor resection in mice. *Int. J. Immunopharmac.*, **16**, 29–36.

Kelkar, S. M., Shenoy, M. S. and Kaklij, G. S. (1988) Antitumour activity of lactic acid bacteria on a solid fibrosarcoma, sarcoma-180 and Ehrlich ascites carcinoma. *Cancer Lett.*, **42**, 73–7.

Kimura, N. T., Yaniguchi, S., Aoki, K. and Baba, T. (1980) Selective localization and growth of *Bifidobacterium bifidum* in mouse tumours following intravenous administration. *Cancer Res.*, **40**, 2061–8.

Kitazawa, H., Itoh, T. and Yamaguchi, T. (1991a) Induction of macrophage cytotoxicity by slime products produced by encapsulated *Lactococcus lactis* ssp. *cremoris*. *Anim. Sci. Technol. (Jpn)*, **62** 861–6.

Kitazawa, H., Nomura, M., Itoh, T. and Yamaguchi, T. (1991b) Functional alteration of macrophages by a slime-forming *Lactococcus lactis* ssp. *cremoris*. *J. Dairy Sci.*, **74**, 2082–8.

Kitazawa, H., Toba, T., Itoh, T. *et al.* (1991c) Antitumoural activity of slime-forming, encapsulated *Lactococcus lactis* subsp. *cremoris* isolated from Scandinavian ropy sour milk, 'villi'. *Anim. Sci. Technol. (Jpn)* **62**, 277–83.

Kitazawa, H., Matsumura, K., Itoh, T. and Yamaguchi, T. (1992) Interferon induction in murine peritoneal macrophage by stimulation with *Lactobacillus acidophilus*. *Microbiol. Immunol.*, **36**, 311–15.

Kitazawa, H., Tomioka, Y., Matsumura, K. *et al.* (1994) Expression of mRNA encoding IFN in macrophages stimulated with *Lactobacillus gasseri*. *FEMS Microbiol. Lett.*, **120**, 315–22.

Kiyosawa, I., Matsuyama, J. and Arai, C. (1992) Antimutagenicity of water and methanol extracts of soymilks fermented by Bifidobacteria and lactic acid bacteria. *Nippon Shokuhin Kogyo Gakkaishi*, **39**, 939–44.

Klaenhammer, T.R. (1993) Genetics of bacteriocins produced by lactic acid bacteria. *FEMS Microbiol. Rev.*, **12**, 39–86.

Kohwi, Y., Hashimoto, Y. and Tamura, Z. (1982) Antitumour and immunological adjuvant effect of *Bifidobacterium infantis* in mice. *Bifidobacteria Microflora*, **1**, 61–8.

Kohwi, Y., Imai, K., Tamura, Z. and Hashimoto, Y. (1978) Antitumour effect of *Bifidobacterium infantis* in mice. *Gann*, **69**, 613–18.

Kuwabara, M., Kosaka, T., Tanaka, S. *et al.* (1988) Effect of *Lactobacillus casei*: formation of interferon and macrophage activating factor in mice *in vivo*. *Jpn J. Vet. Sci.*, **50**, 665–72.

Lichtenstein, A. K., Kahle, J., Berek, J. and Zighelbom, J. (1984) Successful immunotherapy with intraperitoneal *Corynebacterium parvum* in a murine ovarian cancer model is associated with the recruitment of tumour-lytic neutrophils into the peritoneal cavity. *J. Immunol.*, **133**, 519–26.

Lichtenstein, A. H. and Goldin, B. R. (1993) Lactic acid bacteria and intestinal drug and cholesterol, in *Lactic Acid Bacteria* (ed. S. Salminen and A. von Wright), Marcel Dekker, New York, pp. 227–35.

Mann, G. V. and Spoerry, A. (1974) Studies of a surfactant and cholesterolemia in the Massai. *Am. J. Clin. Nutr.*, **27**, 464–9.

Masuno, T., Kishimoto, S., Ogura, T. *et al.* (1991) A comparative trial of LC9018 plus Doxorubicin and Doxorubicin alone for the treatment of malignant pleural effusion secondary to lung cancer. *Cancer*, **68**, 1495–500.

Matsumura, K., Kitazawa, H., Itoh, T. and Yamaguchi, T. (1992) Interferon induction by murine peritoneal macrophage stimulated with *Lactobacillus gasseri*. *Anim. Sci. Technol. (Jpn)*, **63**, 1157–9.

Matsuzaki, T., Yokokura, T. and Azuma, I. (1985) Anti-tumour activity of *Lactobacillus casei* on Lewis lung carcinoma and line-10 hepatoma in syngeneic mice and guinea pigs. *Cancer Immunol. Immunother.*, **20**, 18–22.

Matsuzaki, T., Yokokura, T. and Azuma, I. (1987) Antimetastatic effect of *Lactobacillus casei* YIT9018 (LC9018) on a highly metastatic variant of B16 melanoma in C57BL/6J mice. *Cancer Immunol. Immunother.*, **24**, 99–105.

Matsuzaki, T., Yokokura, T. and Mutai, M. (1988) Antitumour effect of intra-pleural administration of *Lactobacillus casei* in mice. *Cancer Immunol. Immunother.*, **26**, 209–14.

Mehata, A. M., Patel, K. A. and Dave, P. J. (1983) Isolation and purification of an inhibitory protein from *Lactobacillus acidophilus* AC1. *Microbios*, **37**, 37–43.

Metchnikoff, E. (1908) *The Prolongation of Life* (Translation by P. C. Michell), C. P. Putnam's Sons, New York, pp. 161–83.

Montes, R. G., Bayless, T. M., Saavedra, J. M. and Perman, J. A. (1995) Effect of milks inoculated with *Lactobacillus acidophilus* or a yogurt starter culture in lactose-maldigesting children. *J. Dairy Sci.*, **78**, 1657–64.

Moore, W. E. C. and Moore, L. H. (1995) Intestinal floras of populations that have a high risk of colon cancer. *Appl. Environ. Microbiol.*, **61**, 3202–7.

Morotomi, M. and Mutai, M. (1986) *In vitro* binding of potent mutagenic pyrolysates to intestinal bacteria. *J. Natl Cancer Inst.*, **77**, 195–201.

Nadathur, S. R., Gould, S. J. and Bakalinsky, A. T. (1994) Antimutagenicity of fermented milk. *J. Dairy Sci.*, **77**, 3287–95.

Nanno, M., Ohwaki, M. and Mutai, M. (1986) Induction by *Lactobacillus casei* of increase in macrophage colony-forming cells and serum colony-stimulating activity. *Jpn J. Cancer Res. (Gann)*, **77**, 703–10.

Nishioka, K., Miyamoto, T., Kataoka, K. and Nakae, T. (1989) Preliminary studies on antimutagenic activities of lactic acid bacteria. *Jpn J. Zootech. Sci.*, **60**, 491–4.

Norred, E. P. and Voss, K. A. (1994) Toxicity and role of fumonisins in animal diseases and human esophageal cancer. *J. Food Pro.*, **57**, 522–7.

Novick, A. and Szilard, L. (1952) Anti-mutagens. *Nature* (London), **170**, 926–7.

Okawa, T., Kita, M., Arai, T. *et al.* (1989) Phase II randomized clinical trial of LC9018 concurrently used with radiation in the treatment of carcinoma of uterine cervix: its effect on tumour reduction and histology. *Cancer*, **64**, 1769–76.

O'Sullivan, M. G., Thornton, G., O'Sullivan, G. C. and Collins, J. K. (1992) Probiotic bacteria: myth or reality? *Trends Food Sci. Technol.*, **3**, 309–14.

Owen, R. L. and Jones, A. L. (1974) Epithelial cell specialization within human Peyer's patches: an ultrastructural study of intestinal lymphoid follicles. *Gastroenterology*, **66**, 189–203.

Owen, R. L., Pierce, N.F., Apple, R. T. and Cray, W. C. Jr (1986) M cell transport of *Vibrio cholerae* from the intestinal lumen into Peyer's patches: a mechanism for antigen sampling and for microbial transepithelial migration. *J. Infect. Dis.*, **153**, 1108–18.

Perdigon, G., Alvarez, S., Nader de Macias, M. E. *et al.* (1986a) Lactobacilli administered orally induce release of enzymes from peritoneal macrophages in mice. *Milchwissenschaft*, **41**, 344–8.

Perdigon, G., Nader de Macias, M. E., Alvarez, S. *et al.* (1986b) Effect of perorally administered lactobacilli on macrophage activation in mice. *Infect. Immun.*, **53**, 404–10.

Perdigon, G., Nader de Macias, M. E., Alvarez, S. *et al.* (1987) Enhancement of immune response in mice fed with *Streptococcus thermophilus* and *Lactobacillus acidophilus*. *J. Dairy Sci.*, **70**, 919–26.

Popova, P., Guencheva, G., Davidkova, G. *et al.* (1993) Stimulation effect of Deodan (an oral preparation from *Lactobacillus bulgaricus* 'LB51') on monocytes/macrophages and host resistance to experimental infection. *Int. J. Immunopharmac.*, **15**, 25–37.

Reddy, G. V., Shahani, K. M. and Banerjee, M. R. (1973) Inhibitory effect of yogurt on Ehrlich ascites tumour-cell proliferation. *J. Natl Cancer Inst.*, **50**, 815–17.

Reddy, G. V., Friend, B. A., Shahani, K. M. and Farmer, R. E. (1983) Antitumour activity of yogurt components. *J. Food Pro.*, **46**, 8–11.

Robinson, R. K. (1989) Special yogurts: the potential health benefits. *Dairy Ind. Int.*, **54**, 23–5.

Saavedra, J. M. and Perman, J. A. (1989) Current concepts in lactose malabsorption and intolerance. *Annu. Rev. Nutr.*, **9**, 475.

Sakamoto, K. and Konishi, K. (1988) Antitumour effect of normal intestinal microflora on Ehrlich ascites tumour. *Jpn J. Cancer Res. (Gann)*, **79**, 109–16.

Savage, D. C. (1995) Ecology of the intestinal microflora, in *The International Scientific Forum on 'Milk, Dairy Products and Intestinal Microflora'* (Lecture 1), Tokyo (Nov. 25, 1995), pp. 8–9.

Savaiano, D. A. and Levitt, M. D. (1987) Milk intolerance and microbe containing dairy foods. *J. Dairy Sci.*, **70**, 356–97.

Sekine, K., Toida, T., Saito, M. *et al.* (1985) A new morphologically characterized cell wall preparation (whole peptidoglycan) from *Bifidobacterium infantis* with a higher efficacy on the regression of an established tumour in mice. *Cancer Res.*, **45**, 1300–7.

Sekine, K., Watanabe-sekine, E., Ohta, J. *et al.* (1994) Induction and activation of tumouricidal cells *in vitro* and *in vivo* by the bacterial cell wall of *Bifidobacterium infantis*. *Bifidobacteria Microflora*, **13**, 65–77.

Shahani. K. M., Friend, B. A. and Bailey, P. J. (1983) Antitumour activity of fermented colostrum and milk. *J. Food Prot.*, **46**, 385–6.

Shimizu, T., Nomoto, K., Yokokura, T. and Mutai, M. (1987) Role of colony-stimulating activity in antitumour activity of *Lactobacillus casei* in mice. *J. Leukocyte Biol.*, **42**, 204–12.

Silva, M., Jacobus, N. V., Deneke, C. and Gorbach, S. L. (1987) Antimicrobial substance from a human *Lactobacillus* strain. *Antimicrob. Agents Chemother.*, **31**, 1231–3.

Sugimura, T. (1982) Mutagens, carcinogens, and tumour promoters in our dairy food. *Cancer*, **49**, 1970–83.

Suzuki, I., Nomura, M. and Morichi. T. (1991) Isolation of lactic acid bacteria which suppress mold growth and antifungal action. *Milchwissenschaft*, **46**, 640–4.

Takahashi, M., Iwata, S., Yamazaki, N. and Fujiwara, H. (1991) Phagocytosis of the lactic acid bacteria by M cells in the rabbit Peyer's patches. *J. Clin. Electron Microscopy*, **24**, 532–8.

Tanabe, T., Otani, H. and Hosono, A. (1991) Binding of mutagens with cell wall peptidoglycan of *Leuconostoc mesenteroides* subsp. *dextranicum* T-180. *Milchwissenschaft*, **46**, 622–5.

Tanabe, T., Suyama, K. and Hosono, A. (1994) Effect of sodium dodecylsulphate on the binding of *Lactococcus lactis* subsp. *lactis* T-80 cells with Trp-P1. *J. Dairy Res.*, **61**, 311–15.

Thyagaraja, N. and Hosono, A. (1993) Antimutagenicity of lactic acid bacteria from 'Idly' against food-related mutagens. *J. Food Pro.*, **56**, 1061–6.

Thyagaraja, N. and Hosono, A. (1994) Binding properties of lactic acid bacteria from 'Idly' towards food-borne mutagens. *Food and Chem. Toxicol.*, **32**, 805–9.

Toba, T. Yoshioka, E. and Itoh, T. (1991a) Lacticin, a bacteriocin produced by *Lactobacillus delbrueckii* subsp. *lactis. Letters Appl. Microbiol.*, **12**, 43–5.

Toba, T., Yoshioka, E. and Itoh, T. (1991b) Acidophilucin A, a new heat-labile bacteriocin produced by *Lactobacillus acidophilus* LAPT 1060. *Letters Appl. Microbiol.*, **12**, 106–8.

Tsuyuki, S., Yamazaki, S., Akashiba, H. *et al.* (1991) Tumour-suppressive effect of a cell wall preparation, WPG, from *Bifidobacterium infantis* in germfree and flora-bearing mice. *Bifidobacteria Microflora*, **10**, 43–52.

Vincent, J. G., Veomett, R. C. and Riley, R. I. (1959) Antibacterial activity associated with *Lactobacillus acidophilus*. *J. Bacteriol.*, **78**, 477.

Yamaguchi, T., Hoshino, F., Aso, H. *et al.* (1981) *Streptococcus faecalis* as bacterial immunostimulant, in *Current Chemotherapy and Immunotherapy. Proceedings of 12th International Congress of Chemotherapy*, 1175–7. American Society of Microbiology. Washington, DC.

Yamaguchi, T., Kuroda, Y., Saito, M. *et al.* (1984) Immune interferon production by TH69, a lyophilized preparation of *Streptococcus faecalis*, in murine spleen cell culture. *Microbiol. Immunol.*, **28**, 601–10.

Yamazaki, S., Machii, K., Tsuyuki, S. *et al.* (1985) Immunological responses to monoassociated *Bifidobacterium longum* and their relation to prevention of bacterial invention. *Immunology*, **56**, 43–50.

Yasui, H., Mike, A. and Ohwaki, M. (1989) Immunogenicity of *Bifidobacterium breve* and change in antibody production in Peyer's patches after oral administration. *J. Dairy Sci.*, **72**, 30–5.

Yasutake, N., Kato, I., Ohwaki, M. *et al.* (1984) Host-mediated antitumour activity of *Lactobacillus casei* in mice. *Gann*, **75**, 72–80.

Yukuchi, H., Goto, T. and Okonogi, S. (1992) The nutritional and physiological value of fermented milks and lactic milk drinks, in *Functions of Fermented Milk* (eds Y. Nakazawa and A. Hosono), Elsevier Applied Science, London, pp. 2–45.

Zbar, B., Bernstein, I. D., Bartlett, G. L. *et al.* (1972) Immunotherapy of cancer: regression of intradermal tumours and prevention of growth of lymph node metastases after intralesional injection of living *Mycobacterium bovis. J. Natl Cancer Inst.*, **49**, 119–30.

CHAPTER 6

Stimulation of immunity by probiotics

G. Famularo, S. Moretti, S. Marcellini and C. De Simone

6.1 INTRODUCTION

The habit of consuming fermented milk has a long history going back hundreds of years. However, it was not until the end of the last century that the consumption of fermented milk was related to health.

In general, probiotics are regarded as preparations consisting of live micro-organisms or microbial stimulants which affect the endogenous microflora of the recipient (Fuller, 1995). This definition refers to micro-organisms and would include bacteria, yeasts, fungi, viruses and bacteriophages. The micro-organisms included as probiotics are usually assumed to be non-pathogenic components of the normal microflora, such as the lactic-acid-producing bacteria (LAB). However, there is evidence that even non-pathogenic variants of pathogenic bacterial species can operate in much the same way as traditional probiotics. For example, avirulent mutants of *Escherichia coli, Clostridium difficile,* and *Salmonella typhimurium* can also protect against infection by the corresponding virulent parent strains (Fuller, 1995).

The protective effects of the endogenous microflora of the gut are beyond question, as shown by the evidence that certain components of the gut microflora protect the host against infectious diseases (Fuller, 1995). The many claims that have been made for probiotics include suppression of diarrhoea, relief of lactose intolerance, growth stimulation of farm animals and stimulation of mucosal and systemic immunity. However, the results of some reports appear inconsistent and variable and provide a stick with which to beat the probiotic concept. In

Probiotics 2: Applications and practical aspects.
Edited by R. Fuller.
Published in 1997 by Chapman & Hall.   ISBN 0 412 73610 1

many cases these differences are more apparent than real, though factors related to both the micro-organism and the host may affect the results of a probiotic trial. Micro-organism-related factors include the type of micro-organism in the probiotic, the method of production and administration, and the viability of the probiotic preparation. It is known that in commercially available probiotic preparations the counts of viable LAB may vary greatly and sometimes LAB other than the one listed on the label are present. Furthermore, studies designed to assay commercially available probiotic preparations have demonstrated that a few did not contain viable LAB (Gilliland, 1981). Among the host-related factors impacting on the effectiveness of probiotic treatment, the condition of the gut microflora appears to have a critical role.

It is expected that more knowledge of how probiotics work and the optimal methods for administration will enable us to select more active strains and administer them in a way that will make the results more consistent and predictable.

6.2   MICROFLORA, PROBIOTICS AND IMMUNITY

Mucosal surfaces are continually exposed to antigenic substances and the effectiveness of the ensuing mucosal immune response as well as the induction of immune tolerance are regarded as critical in maintaining the health of the host. Probiotics may affect both mucosal and systemic immune responses (Table 6.1). Several reports have focused on the ability of LAB to modulate the production of cytokines, the mitogen- and antigen-driven lymphocyte proliferation, natural killer (NK) cell cytotoxicity, antibody production as well as some metabolic pathways of monocytes–macrophages. There is also mounting evidence that LAB may be involved in the pathogenesis of some models of autoimmunity in experimental animals and, possibly, also in humans. However, the investigation of the immunomodulating properties of

Table 6.1   Effects of probiotics on systemic and mucosal immunity

- Modulation of IL-1, IL-6, TNF-alpha, IL-2, IFNs (gamma, alpha, beta), M-CSF production
- Modulation of antigen- and mitogen-driven mononuclear cell proliferation
- Modulation of monocyte–macrophage phagocytic and killing activities
- Modulation of autoimmunity
- Modulation of immunity to *Salmonella typhi*
- Modulation of immunity to *Cryptosporidium parvum*

IL, interleukin; TNF, tumour necrosis factor; IFNs, interferons; M-CSF, monocyte colony-stimulating factor.

probiotics, and particularly LAB, may prove a hard issue because of the complexity of the systems involved, the gut microflora and the cytokine and cell networks activated during both mucosal and systemic immune responses. Therefore, before summarizing the available knowledge on the stimulation of immunity by probiotics (LAB) it is important to consider the endogenous microflora of the gut as well as the morphology and function of the gut-associated lymphoid tissue (GALT).

6.3 ENDOGENOUS MICROFLORA

LAB are represented among the members of the normal microflora and inhabit the gastrointestinal tract of many animal species, such as humans, pigs, fowl and rodents. Even though many bacterial species of the gut microflora, including some Gram-negative facultative and obligate anaerobes, produce lactic acid during the fermentation of carbohydrates, it is common to restrict reference of LAB to Gram-positive species (*Streptococcus, Lactobacillus, Bifidobacterium* and *Enterococcus*).

Many factors have been shown to affect the prevalence and distribution of LAB in the gastrointestinal tract. In humans it has been shown that the stomach acidity and the propulsive motor activity of the upper small intestine reduce bacterial colonization, thus keeping the upper gut almost sterile (Edwards, 1994). However, the flora of the distal small intestine increases along its length and reaches 10^{10} cfu g^{-1} in the colon. In subjects from developing countries the small intestine has heavier populations of bacteria, which may relate to the higher incidence of diarrhoea and malnutrition (Edwards, 1994). By contrast, in other animal species, such as pigs, fowl and rodents, high counts of LAB may be found throughout the entire gastrointestinal tract and this is probably a consequence of their ability to adhere to, and colonize, the epithelial surface in the proximal regions of the gastrointestinal tract in these animals.

The microbial ecosystem of the lower part of the gastrointestinal tract can be divided into at least two subecosystems: the luminal flora and the mucosal flora. The composition of the luminal flora is mainly determined by the nutrients available and the effects of antimicrobial substances, whereas the composition of the mucosal flora is mainly determined by the host's expression of specific adhesion sites on the enterocyte membrane, the rate of mucus production, the production of secretory immune globulin (Ig) and the extrusion of cellular material from the membrane into the mucus.

The colonizing ability of probiotic bacteria, including LAB, depends mainly on their ability to adhere to the epithelial surface of the gut. There is evidence that adhered micro-organisms do not need to multiply at a high rate and can withstand mucosal washout while maintaining a slow metabolic rate. This probably explains why adhering micro-organisms can occupy niches which are too low in substrates for most other non-adhering micro-organisms (Jansen and Van der Waaij, 1995). By contrast, closer to the lumen, associated micro-organisms withstand washout from the mucus by rapid multiplication without adherence. This group of micro-organisms requires access to an excess of substrates and needs an efficient, probably oxidative, type of metabolism (Jansen and Van der Waaij, 1995).

The bacterial ecosystem in the colon comprises up to 400 species, which are predominantly anaerobic and capable of fermenting carbohydrate and protein as well as metabolizing a wide and diverse range of endogenous and exogenous molecules, including bile acids, fats and drugs (Edwards, 1994). The main substrates for the bacteria of the endogenous microflora are dietary fibre (non-starch polysaccharides), unabsorbed sugars and oligosaccharides, and starch that has escaped digestion in the small intestine (Englyst, Kingman and Cummings, 1992).

The factors driving the initial colonization of the gut are not entirely known, but environment and diet have a crucial role. Babies fed exclusively mother's breast milk have a different flora as compared to babies fed formula derived from cow's milk (Balmer and Wharton, 1989). Bottle-fed babies have faecal flora more typical of an adult. The dietary factors that influence the initial colonization include the buffering capacity of the milk, casein/whey and phosphate content, oligosaccharides, bifidus factor, secretory IgA, lactoferrin, iron, and the presence of nucleotides in breast milk (Edwards, 1994). The differences found in the faecal flora may explain the lower rate of gastrointestinal infections in breast-fed babies compared with formula-fed babies, even after confounding factors such as the social class have been accounted for (Howie *et al.*, 1990). During weaning, the faecal flora develops towards that of the adult flora and bacteria become more numerous and the ecosystem more complex. Furthermore, for breast-fed babies this period of microflora instability is associated with a greater risk of infectious diarrhoea (Edwards, 1994). In turn, the faecal flora of adults is remarkably stable and diet changes have little impact on the endogenous microflora. However, there is evidence that diets rich in fermentable dietary fibres increase bacterial counts and the expression of a variety of enzymes (Edwards, 1994).

Imbalances in the endogenous microflora are frequently observed in several diseases of humans and animals. Many reports have shown that the maintenance of a normal microflora in the gut is crucial for protecting the host against colonization by pathogens (colonization resistance). In other words, a normal balance among bacteria of the indigenous gut microflora prevents any one of a few bacteria becoming dominant and also prevents pathogenic bacteria, which may invade from the environment, from becoming established in the ecosystem of the gut (Freter and Nader de Macias, 1995). Furthermore, pathogenic bacteria entering the lamina propria from the gut lumen may proliferate and translocate to other organs, but a normal endogenous microflora counteracts this (Freter and Nader de Macias, 1995). In turn, the entry of bacteria from the physiological indigenous microflora into the mucosa and their subsequent translocation to other organs is currently regarded as a crucial step for the development of the normal mucosal and systemic immunity.

6.4 MORPHOLOGY OF GUT-ASSOCIATED LYMPHOID TISSUE

In the gastrointestinal tract there are specialized lymphoid aggregates (GALT), which include the tonsils and adenoids, Peyer's patches (PP), appendix and solitary lymphoid nodules. These aggregates share a relatively standard morphology and should be regarded as major inductive sites for gastrointestinal mucosal immune responses. In addition, most are considered to be secondary lymphoid tissues which are responsive to antigens and generate immune responses throughout life.

The gastrointestinal tract contains as much lymphoid tissue as the spleen does. The system can be morphologically and functionally subdivided into two major parts:

1. organized tissues consisting of the mucosal follicles (GALT);
2. a diffuse lymphoid tissue consisting of the widely distributed cells located in the mucosal lamina propria.

The former tissues are 'afferent' lymphoid areas, where antigens enter the system and induce immune responses, and the latter are 'efferent' lymphoid areas, where antigens interact with differentiated cells and cause the secretion of antibodies by B cells or induce cytotoxic reaction by T cells.

The lymphoid aggregates of GALT are morphologically different from lymphoid aggregates of the systemic lymphoid system. More particularly, antigen enters through specialized epithelial cells, called M cells (membranous cells), in the epithelium overlying the lymphoid aggregates. M cells take up materials in the gut lumen and release them in an ungraded form into the subepithelial area. Such transport is applicable to widely disparate substances including particulates (viruses, bacteria and protozoa) and soluble proteins. Remarkably, neither binding nor uptake by M cells is totally indiscriminate.

The area just below the epithelium of the lymphoid aggregates (the so-called dome area) is rich in cells bearing class II major histocompatibility (MHC) antigens (macrophages, dendritic cells and B cells) capable of antigen presentation following *in vitro* exposure to antigens or *in vivo* oral antigen feeding. M cells do not usually bear class II MHC antigens and are, therefore, probably not involved in antigen presentation; however, particularly in the presence of inflammation, they do express class II MHC antigens and may have antigen-presenting function.

Below the dome area is the follicular zone, which contains the germinal centres. B cells predominate in this region although scattered T lymphocytes are also present. The lymphoid aggregates of GALT form the site of IgA B cell development, but the mucosal follicle is conspicuous for the absence of the terminally differentiated IgA B cells (IgA plasma cells) presumably because such cells leave the follicle before differentiating into plasma cells. The interfollicular areas between and around the follicles are also enriched in T cells and most of the small population of CD8 T lymphocytes in mucosal lymphoid aggregates are found in these areas.

The diffuse lymphoid tissues of GALT consist of cell populations present in two separate compartments (the intra-epithelial lymphocyte compartment and the lamina propria lymphocyte compartment).

6.5 INTRA-EPITHELIAL LYMPHOCYTE COMPARTMENT

Within the normal intestinal epithelium there is a large population of leucocytes which are mostly accounted for by lymphocytes (intra-epithelial lymphocytes, IELs). Mucosal lymphocyte movement is regulated by a series of receptors and counter-receptors that are located on lymphocytes and high-endothelial venules. The retention of T cells within the intra-epithelial space is very efficient and stable. Mixing experiments between chimeric animals suggests there is little egress of IELs. It has been recently shown that T cells and mast cells share common homing/adhesion receptors, which suggests that they have evolved to use similar migration pathways. Furthermore, these cells colocalize in the gastrointestinal mucosa, and it is possible that they act in concert to achieve an immune response.

The majority of IELs express the T-lymphocyte marker, CD3, and about 90% have a CD8 phenotype and the remainder a CD4 phenotype. About 20% of IELs express the alpha–beta heterodimeric form of the T-cell receptor for antigen (TCR), whereas about 38% express the TCR gamma–delta form. The marked abundance of CD8 cells and TCR gamma–delta expression are the most prominent characteristics of IELs. Many studies have recently highlighted the significance of cross-talk between IELs and epithelial cells for IEL growth and development (Kagnoff, 1996). For example, intestinal epithelial cells produce interleukin (IL)-7 and stimulate gamma–delta IELs to proliferate. Furthermore, the expression of IL-7 receptor is upregulated on gamma-delta IELs following the stimulation with IL-7 or IL-2. In turn, IL-7 'knock-out' mice lack gamma–delta IELs, whereas the alpha–beta population is only slightly reduced (Kagnoff, 1996).

The spectrum of functions of IELs is not completely known, but many of these cells exert strong cytotoxic activity (Cerf-Bensussan and Guy-Grand, 1991). In addition, IELs contain NK cells, mast cells and T-cell precursors and can produce interferon (IFN)-gamma, IL-2 and IL-5 (Befus, 1993). IELs may also be involved in the suppression of mucosal hypersensitivity reactions as well as in contrasuppressor reactions (Fujihashi *et al.*, 1990; Perdue and McKay, 1993). Recent evidence proposed that gamma–delta IELs can regulate the mucosal IgA response and this concept gained further support from studies in gamma–delta 'knock-out' mice that demonstrated a marked decrease in IgA plasma cells and decreased IgA responses to antigen stimulation (Kagnoff, 1996). By contrast, IgG and IgM responses are normal in this model.

The frequency of gamma–delta T lymphocytes among IELs increases considerably during coeliac disease and protozoan infections, such as *Giardia lamblia* (Perdue and McKay, 1993). Furthermore, gamma–delta T lymphocytes may enhance the expression of MHC class II antigens upon enterocyte surface via the production of IFN-gamma. Because IELs are found in close relationship with the basement membrane and may be in transit between an intra-epithelial niche and a submucosal location, they have the potential to influence the functions of epithelia of both the compartments.

6.6 LAMINA PROPRIA LYMPHOCYTE COMPARTMENT

The lymphocyte population beneath the epithelial layer in the lamina propria, the LPL population, is about equally divided between T and B cells. The B-lymphocyte population is dominated by IgA B lymphocytes and plasma cells. In both normal and diseased mucosal tissue, the B-cell population is composed of cells that under *in vitro* conditions display spontaneous immunoglobulin secretion. The LPL T-lymphocyte population is composed of both CD4 and CD8 lymphocytes that have undergone prior activation. Recent studies have led to the concept that LPL T cells are a class of memory T lymphocytes.

6.7 ANTIGEN PRESENTATION IN THE GUT

An important route of antigen entry to mucosal surfaces of the gut involves the M cells, which are located in follicle-associated epithelium. Remarkably, M cells have recently received considerable attention because of the potential to target vaccine vectors to the mucosal lymphoid compartments.

The intestinal uptake of antigens by epithelial M cells emphasizes the association between immune response, including the production of antibodies, mostly IgA, and the mucosal epithelium (Perdue and McKay, 1993). In contrast to the intestinal epithelial cells that take up soluble antigens, M cells are more efficient at binding and transporting particulate antigens, and their surface appears to be specialized for endocytosis. In face, M cells overlay GALT and have a highly invaginated basolateral membrane in which antigen-presenting cells (APCs), such as macrophages and dendritic cells, and lymphocytes reside. In this location, antigens are taken up and processed by APCs and then presented in conjunction with products of MHC class II genes to lymphocytes for triggering the immune response.

Notably, in both animals and humans the intestinal mucosa contains an excess of macrophages relative to dendritic cells (Mayrhofer, 1994). This excess has been suggested to function to suppress the level of antigen presentation to local T lymphocytes in the lamina propria (Mayrhofer, 1994). In PP, as well as other GALT structures, the numerical predominance of dendritic cells over macrophages may favour effective presentation of antigens (Mayrhofer, 1994). The function of macrophages is related more to the antibacterial defences of PP, which are at high risk because of the ability of M cells to transport potential pathogens, whereas the antigen-presenting activity is mostly performed by dendritic cells which appear as classic MHC class II-positive interdigitating reticulum cells (Mayrhofer, 1994). Furthermore, the enterocytes of intestinal mucosa may also act as 'non-professional' APCs. In fact, in addition to presenting viral antigens to cytotoxic T lymphocytes, enterocytes can also process and present exogenous antigens. This possibility has been raised by the finding that epithelia can express MHC class II antigens either constitutively or in response to certain physiological or local immunological stimuli, such as the presence of cytokines, mostly IFN-gamma, in the microenvironment (Perdue and McKay, 1993).

The epithelial cell is now recognized as a central player for regulating the natural and acquired immune system of the host at mucosal surfaces of the gut.

The view that enterocytes could act as 'non-professional' APCs has been further supported by the finding that enterocytes are also active in the uptake and transepithelial transport of antigens (Mayrhofer, 1994). Finally, the demonstration that intestinal epithelial cells produce cytokines, such as IL-1 and IL-6 which exert costimulatory effects on T-lymphocyte activation, is in agreement with the hypothesis that enterocytes may be involved in antigen presentation (Bland, 1987; Santos *et al.*, 1990).

Intestinal epithelial cells have been shown to provide early signals that are important for the initiation and regulation of the mucosal inflammatory response of the host following bacterial invasion and, in this regard, these cells can be viewed as early sensors of bacterial invasion at mucosal surfaces, according to the demonstration that they upregulate the expression and production of a broad array of pro-inflammatory cytokines and prostanoids in response to bacterial invasion (Kagnoff, 1996). Furthermore, in addition to expressing cytokines, intestinal epithelial cells express several cytokine receptors, such as that for transforming growth factor (TGF)-beta (Kagnoff, 1996). Interestingly, TGF-beta has been shown to play a major role in the restitution of epithelial cells following injury (Kagnoff, 1996).

In addition to the basolateral membranes, MHC class II molecules are expressed also on the brush borders of enterocytes indicating that MHC class II antigens are exposed directly to the human contents of small intestine as well as being expressed on the ablumenal membranes of enterocytes (Mayrhofer, 1994). This could be important to the function and pathology of the gastrointestinal tract since the luminal digestion in the gut could produce immunogenic peptides. Therefore, potentially immunogenic peptides could bind directly to the peptide-binding groove of the MHC molecules. Furthermore, it is conceivable that the peptide-laden molecules could be relocated to the basolateral cell membranes where they would be accessible to mucosal T lymphocytes.

6.8   EFFECTOR RESPONSE IN THE GUT

The involvement of mucosal lymphoid tissue in host defence mechanisms has been extensively studied over the past two decades. GALT is the pivotal site for the induction of mucosal immune response in the gut, including the generation of specific hyporesponsiveness to mucosal antigens (oral tolerance) (McGhee *et al.*, 1989).

Both Th1 and Th2 helper functional CD4 subsets are found in GALT, with their cytokine repertoire. IL-2 and IFN-gamma-producing Th1 cells predominate in lymphoid aggregates whereas Th2 lymphocytes, producing mostly IL-4, IL-5, and IL-6, have been shown to predominate in the mucosal lamina propria where effector functions are foremost (Husband and Dunkley, 1990).

There is also evidence that CD8 T lymphocytes of GALT can generate antigen-dependent cytotoxic effector cells (London *et al.*, 1987) as well as suppressor populations involved, for example, in mediating the systemic tolerance to orally derived antigens (McGhee *et al.*, 1989).

It is worth noting that effector cells whose differentiation and maturation are initiated in GALT, such as IgA-B cells and cytotoxic T lymphocytes, do not complete their maturation in these sites but migrate through lymphatics to the draining mesenteric lymph nodes and subsequently into the thoracic duct and systemic circulation. Then, many of them selectively re-enter and are retained in the intestinal lamina propria where they may proliferate and complete their differentiation to fully functional IgA-producing plasma cells or other effector cells (Befus, 1993).

The concept of humoral immune response at mucosal sites is now established. Mucosal antibody responses include a striking IgA response, resulting from the abundance of IgA precursor cells in GALT (McGhee *et al.*, 1989). However, the mechanisms driving the local induction of mucosa-associated IgA responses remain to be fully understood. Notably, other antibody isotypes are produced locally in the mucosa or are derived from the circulation due to the leakage of plasma proteins into tissue microenvironments. The protective role of secretory antibody at the external mucosal surfaces has been extensively demonstrated for viral infections, but even after infection with enteric bacteria (*E. coli, Shigella, Salmonella, Staphylococcus*) Ig that specifically react with the pathogenic bacteria have been detected in faecal specimens even though their role in protection against reinfection is unclear (Abraham and Ogra, 1994). Secretory IgA may lyse bacteria in the presence of lysozyme and complement and opsonize bacteria and interfere with the adherence of bacterial antigens to mucosal surfaces, thereby limiting bacterial colonization and enhancing the elimination of bacteria (Abraham and Ogra, 1994). Studies on *S. mutans, V. cholerae,* and the enterotoxin of *E. coli* have shown that secretory IgA blocks the binding sites on the bacterial cell wall and prevents the attachment of bacteria or bacterial toxin to the specific receptors on mucosal cell membranes (Abraham and Ogra, 1994). Relatively little work has been done on the role of cell-mediated reactions at various secretory sites. Even though these reactions may be essential in resistance to viral, bacterial and parasitic infections, there is evidence that functional T lymphocytes may not be a major factor in recovery from rotavirus infections. In turn, specific rotavirus antibody response is critical to complete recovery from virus replication and shedding. However, cell-mediated reactions at various mucosal sites could be important in recovery from certain infections once the mucosal epithelium is colonized.

The outcome of immune responses in the gut may be affected by changes in the intestinal microenvironment induced by several factors, including infections, malnutrition, imbalances in endogenous microflora and mucosal injury. The effects of these changes can be viewed as a cascade of events involving modifications of the antigen and its uptake, and altered serum, secretory and cell-mediated immune responses. Under these conditions the intestine may become more permeable to the absorption of dietary and other environmental antigens. For example, there is evidence that malnutrition causes atrophy of the intestine, leading to increased antigen uptake. The combination of mucosal infections and protein deficiency appears to intensify mucosal atrophy and disruption of microvilli; furthermore, acute enteric infection aggravates nutritional deficiencies and, in turn, malnutrition increases the severity of diarrhoea.

Recent work has demonstrated that cells derived from the GALT and, although to a lesser degree, cells from systemic lymphoid sites may localize at other mucosal surfaces, such as in the bronchial mucosa or urogenital tract (Kagnoff, 1996). Also, cells which have been induced at these other mucosal surfaces may localize in the intestinal mucosa. This reciprocity of migration involves, in addition to IgA B cells and other lymphocyte populations, IgA antibodies which are selectively transported from circulation into secretions at all mucosal sites by the receptor–transporter secretory component produced by mucosal epithelial cells. Here IgA antibodies can bind antigens and target them for phagocytosis, thus minimizing antigen entry into the interstitial microenvironment and consequently reducing inflammatory reactions potentially harmful to the tissues. It is worth noting that the function of IgA antibody at mucosal surfaces appears more extensive than previously recognized. For example, viruses can be neutralized within the epithelial cell during their intracellular life cycle by being trapped during its transport from the basolateral to apical surface (Kagnoff, 1996). Overall, these data point out that the mucosa-associated immune system is a functional system integrated by cell and molecule exchange among all mucosal surfaces throughout the body.

6.9   MODULATION OF GUT FUNCTIONS BY IMMUNE NETWORKS

The induction of local immune response in the gut may in turn affect the secretory and absorbtive functions of enterocytes as well as the motility of the gut.

Cytokines produced by IELs modulate both directly and indirectly, via the recruitment of inflammatory cells, several functions of enterocytes. IL-2 may directly or indirectly, through inducing the release of proinflammatory cytokines IL-1 and tumor necrosis factor (TNF)-alpha, cause the secretion of electrolytes and water into the human intestine (Rosenberg, Lotze and Mule, 1988; Perdue and McKay, 1993). Furthermore, TNF-alpha modulates the expression of the secretory component of IgA antibodies, thereby regulating the transport of IgA across the epithelium (Kvale *et al.*, 1988). In turn, IFN-gamma significantly reduces the capability of epithelial cells to respond to VIP (vasoactive intestinal peptide), cholera toxin and cholinergic agonists, such as carbachol (Holmgren *et al.*, 1989). IFN-gamma also induces the expression of

MHC class II antigens upon enterocyte surfaces, as reported above (Perdue and McKay, 1993), thereby contributing to the amplification of local immune response and induction of autoimmunity towards the epithelium. Furthermore, both IFN-gamma and TNF-alpha released by mucosal T cells can directly mediate the killing of epithelial cells (Deem, Shanahan and Targan, 1991).

Immune reactions can activate enterocytes in terms of cell mitosis and cytokine expression, For example, following inflammatory stimuli, Paneth cells can produce TNF-alpha (Keshav *et al.*, 1990) and IL-6 has been found in gastric and intestinal epithelial cells (Shirota *et al.*, 1990) and, in this latter case, cancer cells have been shown to produce more IL-6 than their normal counterparts. Furthermore, activated T lymphocytes from lamina propria can enhance the proliferation rate of intestinal epithelial cells (Ferreira *et al.*, 1990).

Cytokines and immune cells can also affect the functions of mucus-containing goblet cells which are interspersed between columnar epithelial cells. In fact, early reports have suggested that the hyperplasia of goblet cells induced by *Nippostrongylus brasilensis* is dependent on T cells (Perdue and McKay, 1993) and chronic IL-1 administration has been shown to induce mucus release and goblet cell depletion in the murine colon (Butler *et al.*, 1988). The finding that mucus production is enhanced following secondary antigen exposure further supports the view that immune cells are involved in the regulation of this epithelial function (Perdue and McKay, 1993).

The interactions of cytokines and immune cells with smooth muscle, nerves and hormone-secreting cells result in the altered motility of inflamed or sensitized gut (Collins, 1993). In addition, cytokines released during the inflammatory response drive mesenchymal cells (fibroblasts, smooth muscle cells and endothelial cells) to produce eicosanoids, other cytokines, chemotactic and growth factors (Perdue and McKay, 1993). Notably, another important amplifying mechanism of the intestinal inflammatory response is the induction of gene expression for endothelial and macrophage adhesion molecules which act in synergy with other chemoattractants to increase the infiltration of granulocytes, monocytes and lymphocytes into the inflammatory focus (Gundel and Letts, 1994).

6.10 MODULATION OF CYTOKINE PRODUCTION BY PROBIOTICS

The indigenous bacterial microflora modulates the cytokine network which regulates the immune response and drives effector arms towards invading pathogens. For example, peritoneal macrophages from normal mice produced significantly more IL-1 and IL-6 under *in vitro* conditions than those of germ-free mice (Nicaise *et al.*, 1993), but IL-1 and IL-

6 production from germ-free mice implanted with *E. coli* was comparable to normal mice (Nicaise *et al.*, 1993). In turn, *Bifidobacterium bifidum* did not increase the production of these two cytokines (Nicaise *et al.*, 1993). In these experiments, TNF-alpha was produced only by peritoneal macrophages from normal mice and germ-free mice implanted with *E. coli* (Nicaise *et al.*, 1993).

Taken together, these data suggest that Gram-negative bacteria are the most efficient stimulus for driving the production of macrophage-derived cytokines. Also, we should note that the bacterial flora stimulated cytokine production soon after implantation. In addition, recent reports have demonstrated that *L. acidophilus* induces the production of IFN-alpha/beta by murine peritoneal macrophages (Kitazawa *et al.*, 1992, 1994). This suggests that the inducing activity of IFNs may be one of the available biological parameters for designating the dairy products containing *L. acidophilus* as physiologically functional foods. Both of the bacteria commonly found in yoghurt (*L. delbrueckii* ssp. *bulgaricus* and *S. salivarius* ssp. *thermophilus*) have been shown to induce the production of IL-1 beta, TNF-alpha, and IFN-gamma, but not of IFN-alpha and IL-2, by peripheral blood mononuclear cells (PBMCs) from humans (Pereyra *et al.*, 1993). Furthermore, the walls from these bacteria, but not their cytoplasm, induced a comparable cytokine production (Pereyra *et al.*, 1993). These cytokines were also induced by *L. casei, L. acidophilus, Bifidobacterium* spp., and, to a lesser extent, *L. helveticus* (Pereyra *et al.*, 1993). In this report the IFN production was estimated by the 2-5 synthetase activity from PBMCs following a single ingestion of bacteria in yoghurt or sterile milk and the activity of the yoghurt group enzyme was about 80% higher than that of the milk group. However, no cytokine was detectable in the serum (Pereyra *et al.*, 1993). Interestingly, this increase in IFN-gamma production by lymphocytes has been shown to be associated with increased levels of ionized calcium in the serum.

Furthermore, supplementing the diet of ageing mice with live *L. delbrueckii* ssp. *bulgaricus* and *S. salivarius* ssp. *thermophilus* has been shown to completely restore the levels of IFN-gamma and IFN-alpha with respect to control animals (Muscettola *et al.*, 1994). An age-related decline in the production of cytokines, including IFNs, is common and these results suggest a role for probiotics in preventing the decline of immune functions during ageing. The finding in this study that many LAB adhered selectively to the luminal surface of M cells of PPs suggests that there is translocation of bacteria to the underlying lymphoid tissue. In fact, LAB have recently been observed in mesenteric lymph nodes after microbial colonization in germ-free mice (Umesaki *et al.*, 1993), resulting in an efficient mucosal immune response against potentially harmful intestinal microflora and enteropathogenic microorganisms. Another recent report has demonstrated that sphingomyelin, the active substance of kefir, enhances the production of IFN-beta by a human osteosarcoma cell line (Osada *et al.*,1994). There is also evidence that LAB, mostly *S. salivarius* ssp. *thermophilus*, may induce the production of the pro-inflammatory cytokines IL-1-beta, Il-6 and TNF-alpha (Aattouri and Lemonnier, 1995). Furthermore, an oral preparation from *L. delbrueckii* ssp. *bulgaricus* LB51 stimulates the production of both cell-surface-associated and cytoplasmic IL-1 and TNF-alpha (Popova *et al.*, 1993). In addition, an increased activity of macrophage colony stimulating factor (M-CSF) was detected in serum and an increased level of M-CSF mRNA was observed in the livers of mice treated with a heat-killed *L. casei* preparation during experiments of radioprotection (Tsuneoka *et al.*, 1994).

In a recent report from our laboratory, we have shown that the addition of small quantities of yoghurt containing live *L. delbrueckii* ssp. *bulgaricus* and *S. salivarius* ssp. *thermophilus* to concanavalin A (ConA)-driven human PBMCs resulted in a strong enhancement of IFN-gamma production (De Simone *et al.*, 1993). Furthermore, these supernatants augmented NK cell cytotoxicity against K562 targets much more than with respect to control supernatants from PBMC cultures stimulated with ConA only (De Simone *et al.*, 1993). Similar results were found using *L. acidophilus, L. casei* and *L. plantarum* (De Simone *et al.*, 1986). Our hypothesis was that LAB stimulate PBMCs to produce cytokines, such as IL-1 and IL-2, which activate resting NK cells to synthesize and release IFN-gamma, proliferate and exert cytotoxicity. This hypothesis has been further supported by experiments demonstrating the binding of LAB to both CD4 and CD8 cells (De Simone *et al.*, 1988a, 1993). As seen in the case of *Salmonella*-stimulated lymphocytes (De Simone *et al.*, 1986, 1988a; Antonaci and Jirillo, 1985), the binding of LAB to T lymphocytes should be referred to as a potent stimulus for immune cell activation.

Even the administration of LAB under *in vivo* conditions may strongly enhance the production of IFN-gamma. In a recent study, healthy volunteers have received lyophilized dietary lactobacilli (3×10^{12} micro-orgranisms) and 200 g of plain yoghurt at 24-h intervals for 28 days. The control group received skimmed milk in a quantity calorically equivalent to that of the yoghurt group. The results were that a strong increase in the serum levels of IFN-gamma as well as the expansion of both B lymphocytes and NK cells can be attained in the normal host by feeding large quantities of dietary LAB (De Simone *et al.*, 1993).

However, mechanisms other than the increased production of IFN-gamma are involved in mediating the increased resistance toward pathogens induced by LAB. This claim has been demonstrated in an experimental model of *Cryptosporidium parvum* infection (De Simone *et al.*, 1995).

Cr. parvum is a protozoan parasite that causes diarrhoeal disease in a variety of mammals, including humans and economically important livestock species (Fayer, Speer and Dubey, 1990). The disease is especially severe in immunocompromised hosts and has become a major cause of morbidity and mortality among patients with the acquired immunodeficiency syndrome (AIDS) (Ungar, 1990). Mechanisms of immunity to *Cr. parvum* are not well understood (Zu *et al.*, 1992), but several *in vivo* studies suggest that both CD4 lymphocytes and IFN-gamma are critical in resistance and recovery from *Cr. parvum* infection (Ungar *et al.*, 1991; Chen, Harp and Harmsen, 1993a, 1993b), in addition to indigenous bacterial intestinal microflora. In fact, germ-free adult mice are more susceptible to the primary challenge than normal mice and, while severe combined immunodeficient (SCID) mice are relatively resistant to *Cr. parvum* infection, germ-free SCID mice are highly susceptible (Harp *et al.*, 1992). Therefore, the presence of intestinal microflora appears to strongly influence the susceptibility of mice to *Cr. parvum* infection.

There is evidence that the colonization of the gut of germ-free mice with LAB can protect the animals from *Cr. parvum* infection (Harp *et al.*, unpublished). In fact, germ-free mice colonized with LAB were clearly less heavily infected with *Cr. parvum* than controls without LAB (Harp *et al.*, unpublished). However, this protection was not directly correlated with induction of IFN-gamma by LAB (Harp *et al.*, unpublished) since the two groups challenged with *Cr. parvum* both produced mRNA for IFN-gamma despite the fact that mice colonized with LAB were protected from infection and the non-colonized group was not (Harp *et al.*, unpublished). In addition, in these experiments germ-free mice treated with LAB only did not produce message for IFN-gamma (Harp *et al.*, unpublished). Therefore, the colonization of germ-free mice with LAB must be protecting either via some mechanisms not involving IFN-gamma or through an indirect pathway. Additionally, it is possible that IFN-gamma message and/or protein was induced earlier in animals receiving LAB only and disappeared by the time of necropsy while the message persisted in mice injected with both LAB and *Cr. parvum*. It is possible that the group injected with *Cr. parvum* only was not protected from challenge because the induction of IFN-gamma seen in this group was a late event occurring as a result of infection and was not sufficient to prevent colonization with the parasite.

The implications of these findings in SCID mice with experimental *Cr. parvum* infection for other species, such as humans with AIDS, are still not clear. It is interesting that the apparent age-related susceptibility and development of resistance in calves and immunocompetent mice correlates with the acquisition of intestinal microflora (Harp, Woodmansee and Moon, 1990). Similarly, one may speculate that the immunocompromised state of AIDS patients, often coupled with poor nutritional status and extensive antibiotic therapy, may result in an altered intestinal microflora which then contributes to the increased susceptibility to *Cr. parvum* infection. In addition, preliminary data suggest that the treatment of *Cr. parvum*-infected AIDS patients with LAB may be of some benefit in alleviating symptoms (De Simone *et al.*, 1995).

6.11   MODULATION OF GUT-ASSOCIATED LYMPHOID TISSUE BY PROBIOTICS

The physiological indigenous microflora has a key role in influencing the appropriate development of the mucosal immune system (De Simone *et al.*, 1993). It has been suggested that this effect of microflora may be mediated by the adhesion of bacteria to gut-surface epithelium, which results in the active stimulation of GALT.

In this regard, LAB have been shown to modulate several functions of GALT. In fact, PP cell suspension cultures from Balb/c mice fed yoghurt containing living LAB exhibited a strong increase of blasto-genic proliferative responses to mitogens such as phytohaemoagglutinin (PHA) and lipopolysaccharide (LPS) compared with controls (De Simone *et al.*, 1993). There is also evidence that the supplementation with LAB may abolish the suppression of lymphocyte responsiveness to mitogens seen in tumour-bearing mice. Furthermore, the increased cell proliferation to LPS, which is mainly a mitogen for B lymphocytes, was correlated with an expansion of the B-lymphocyte pool in PP (De Simone *et al.*, 1993). These latter data have been further supported by a recent report demonstrating that the content of immunoglobulin-syn-thesizing cells in the jeujunal lamina propria of germ-free mice was significantly increased following oral and intraperitoneal administration of killed *L. acidophilus* strains (Smeyanov *et al.*, 1992).

Results comparable with those seen in PP have been obtained when the mitogen-driven splenocyte proliferation was assayed. Mice fed living LAB had a strong increase of splenocyte proliferation to lectins, including both mitogens for T cells (PHA, ConA) and B cells (PWM) (De Simone *et al.*, 1993). These effects were correlated with an expansion of the T-lymphocyte pool in the spleen (De Simone *et al.*, 1993). Remarkably, a brief treatment with heat proved sufficient to strongly reduce the effects of yoghurt, confirming that the immunomo-dulating properties of probiotics are strictly dependent on the presence of viable LAB (De Simone *et al.*, 1993).

Available data also indicate that LAB which can persist in the gastro-intestinal tract may act as adjuvants to the humoral immune response. Indeed, volunteers who consumed a fermented milk containing *L. acido-philus* and bifidobacteria over a period of 3 weeks during which an attenuated *Sal. typhi* was administered to mimic an enteropathogenic infection had a significant increase in the titre of specific serum IgA in comparison to the control group (Link-Amster *et al.*, 1994). An increase in total serum IgA was also observed. Remarkably, these changes correlated with the increase in *L. acidophilus* and bifidobacterial counts in samples of faecal flora (Link-Amster *et al.*, 1994). Even though the mechanisms accounting for these effects of LAB on humoral immune response remain to be fully established, the increased production of IFN-gamma and the enhancement of macrophage functions (see below) following the administration of LAB probably play a critical role. Furthermore, it is known that IFN-gamma enhances the expression of the secretory component, thus substantially contributing to increase the external transport of dimeric IgA.

6.12   MODULATION OF MACROPHAGE FUNCTIONS BY PROBIOTICS

Mice fed living LAB had an increased production of oxygen metabolites by splenocytes following zymosan stimulation (De Simone *et al.*, 1988b, 1993). In addition, the oral and intraperitoneal administration of *L. acidophilus* to germ-free mice led to a significant rise in the level of luminol-dependent chemiluminescence of peritoneal macrophages (Smeyanov *et al.*, 1992). These data are in agreement with the hypothesis that probiotics containing LAB enhance both the phagocytic activity and the respiratory burst of monocyte–macrophage cells, probably resulting in enhanced phagocytosis and killing of pathogens. Moreover, the possibility that lactic-acid-producing bacteria upmodulate the functions of monocytes–macrophages as APCs should be considered. In this respect, the *in vitro* macrophage- and granulocyte-phagocytosis of *E. coli* was enhanced in healthy volunteers given a fermented milk product supplemented with LAB (Schiffrin *et al.*, 1995). This enhancement was coincidental with faecal colonization by LAB and persisted for 6 weeks after ingestion of the fermented products ceased. By this time, the faecal lactobacilli and bifidobacteria had returned to the concentrations prior to consumption. The increase in phagocytic activity persisted for some time after colonization returned to the original values (Schiffrin *et al.*, 1995). The mechanisms accounting for this effect of LAB remain to be fully elucidated, but it is likely that the enhanced production of IFNs driven by probiotics (see above) may at least in part explain it.

Some probiotic products not only have the capacity to induce the endogenous production of cytokines, but they probably also affect other macrophage and granulocyte functions. Whether LAB modulate the expression of cell-surface molecules that are involved in bacterial uptake by leucocytes, such as integrins and Fc gamma receptor, should be carefully investigated. Alternatively, a serum factor, such as an opsonin, may also play a role in mediating this biological effect. Moreover under *in vitro* conditions LAB can alter macrophage functions with a strain-dependent pattern. For example, the phagocytosis of viable *Sal. typhimurium* by murine macrophage was increased by the pretreatment of the phagocytes with cell-free extracts of *L. acidophilus* and *Bif. longum*, but treatment with *Bif. longum* induced higher phagocytic activity (Hatcher and Lambrecht, 1993).

6.13 MODULATION OF RESISTANCE TOWARDS *SALMONELLA TYPHIMURIUM* BY PROBIOTICS

Under *in vitro* conditions, LAB inhibit the growth of food-borne pathogens, including *Salmonella typhimurium* (Gilliland and Speck, 1977), and this prompted studies to evaluate the mechanisms accounting for this antibacterial effect.

Natural antibiotics synthesized by LAB have been identified and, notably, bulgarican, which is produced by *L. delbrueckii* ssp. *bulgaricus*, has been shown to possess a wide spectrum of *in vitro* antibacterial activity (Davidson and Hoover, 1993). In addition, live microbial therapy has been shown in some reports to be more effective than the administration of antibiotics for treating infections from *Salmonella* spp. (Hitchins *et al.*, 1985). However, these studies were not designed to assay whether the protective effects of LAB against *Salmonella* infections are due to the enhancement of specific immune response.

Protection towards *Salmonella* is mediated in the early phase of infection by macrophages and by specific immunity in the late phase of infection (Akeda *et al.*, 1981; Dichelte, Kaspereit and Sedlacek, 1984).

There is evidence that yoghurt containing viable LAB strongly enhances murine defences against *Sal. typhimurium* through several mechanisms (De Simone *et al.*, 1988c, 1993), as shown by:

- increased antibacterial activity against *Sal. typhimurium* of PP mononuclear cells. Furthermore, this effect seems to be mediated, at least in part, by IgA antibodies. This is pivotal for inducing host resistance towards invading salmonellae since their virulence is related to the ability to survive and multiply within the PP microenvironment (Dichelte, Kaspereit and Sedlacek, 1984);
- strongly increased absolute numbers of phagocytosing macrophages, probably due to the accumulation of migratory macrophages from the pool of circulating monocytes at the sites of infection;
- strongly increased proliferative responses of splenocytes to both T-cell and B-cell mitogens, such as ConA and LPS.

Overall, these immunomodulating effects of administering viable LAB resulted in a strong reduction of *Sal. typhimurium* growth in both spleen and liver, therefore accounting for the higher survival rate of animals treated with probiotics.

6.14 MODULATION OF AUTOIMMUNITY BY PROBIOTICS

There is growing knowledge that the composition of the endogenous intestinal microflora may have an important role in the expression of systemic autoimmunity in both humans and animal models. Bacterial products may cause polyclonal B-cell activation but may even induce antigen-driven production of anti-ds-DNA antibodies on the basis of cross-reactivity between bacterial and human DNA. The bacteria involved might originate from the intestinal tract according to the finding that the DNA component of circulating immune complexes of patients with systemic lupus erythematosus (SLE) hybridized with the *lac* gene of *E. coli* (Terada, Okuhara and Kawarada, 1991). The possibility that antibacterial antibodies are sequestrated in immune complexes from SLE patients has been suggested and this might indicate a possible role for antibacterial antibodies in exacerbations of SLE (Apperloo-Renkema *et al.*, 1995). The observation that the colonization resistance, regarded as the defence capacity of the endogenous microflora against colonization of the gut by foreign bacteria, is reduced in SLE patients indirectly supports this hypothesis (Apperloo-Renkema *et al.*, 1994). Indeed, this could indicate that in SLE more and different bacteria translocate across the gut wall, resulting in a higher chance of subsequent DNA-cross-reacting antibacterial antibody production.

In addition to the SLE model, other evidence suggests that the intraluminal microflora is a pivotal factor even in the inflammatory process of Crohn's disease (Dumonceau *et al.*, 1994): diversion of the faecal stream after ileal resection can prevent recurrence (Rutgeerts *et al.*, 1991) and lowering luminal bacterial concentrations improves active disease (Wellmann *et al.*, 1986). The deleterious action of bacteria could be due to a few of their products exhibiting potent pro-inflammatory activity, such as LPS and FMLP (N-formyl-methionyl-leucyl-phenylalanine). LPS and FMLP leak across the intestinal epithelium of patients with inflammatory bowel disease during active disease: endotoxaemia is commonly seen in subjects with Crohn's disease and this could be responsible for priming of lymphocytes observed in active Crohn's disease (Rutgeerts *et al.*, 1991). When primed lymphocytes come in contact with bacterial products, inflammatory mediators are released in excess and result in intestinal damage. In this respect, one should note that certain broad-spectrum antibiotics prevent chronic colitis, thus reinforcing the view that the maintenance of a normal colonic flora may play a critical role in slowing down the progression of inflammatory gut lesions to chronicity. Interestingly, a chronic granulomatous enteritis with systemic manifestations similar to those observed in Crohn's disease may be produced by intramural injections of peptidoglycan–polysaccharide in Lewis rats (Dumonceau *et al.*, 1994). It can be speculated that qualitative and/or quantitative alterations in gut microflora and the associated increase in intestinal permeability is a primary defect in Crohn's disease and could allow passage of bacterial products into the organism, so initiating an immune response with intestinal damage and systemic manifestations. This idea appears to be in agreement with experimental findings and the failure to identify any single aetiologic agent in Crohn's disease after more than 80 years.

Early reports supported the hypothesis that bacterial overgrowth in the small intestine has a role in the pathogenesis of rheumatoid arthritis (RA). Furthermore, a high frequency or high numbers of *Clostridium perfringens* as well as an association between high counts and active disease have been shown, even though available studies are somewhat conflicting (Struthers, 1986; Shinebaum *et al.*, 1987). Recently, the improvement in RA disease activity following the treatment with sulphasalazine has been correlated with decreased counts of anaerobic bacteria, including *C. perfringens*, and *E. coli* (Kanerud *et al.*, 1994). One can only speculate whether the observed effect of sulphasalazine on the faecal flora in RA patients influences disease activity. However, in experimental models of arthritis, the manipulation of gut microflora has been shown to modulate the intensity of the disease (Thompson and Elson, 1993). Although further studies are needed, the confirmation of the above hypothesis would support the use of oral bacteriotherapy with LAB in the treatment of active RA.

The induction of coronary arteritis in mice by *L. casei* cell wall is thought to represent an animal model of the Kawasaki disease (Tomita, Myones and Shulman, 1993). In fact, under *in vitro* conditions the treatment of vascular endothelial cells with supernatants from human PBMCs stimulated with *L. casei* cell wall has been shown to both enhance the adherence of polymorphonuclear cells (PMNs) to human endothelial cells and increase the expression of intercellular adhesion molecule-1 (ICAM-1) (Tomita, Myones and Shulman, 1993). Supernatants contained high concentrations of TNF-alpha and PMN adherence correlated directly with the concentration of TNF-alpha and both ICAM-1 expression and enhanced PMN adherence were inhibited by anti-TNF-alpha treatment (Tomita, Myones and Shulman, 1993). In addition, the initial coronary inflammatory reaction in the mouse model has been shown to involve PMN adherence to vascular endothelium that has been activated by TNF-alpha released by PBMCs following stimulation with *L. casei* cell wall (Tomita, Myones and Shulman, 1993).

6.15 ROLE OF PROBIOTICS IN THE TREATMENT OF HUMAN INFECTIONS

The ability of probiotics to affect both systemic and mucosa-associated immune responses suggests that the administration of these preparations could strongly contribute to promote the recovery from infections (Table 6.2).

Rotavirus infection is a leading cause of infantile diarrhoea worldwide. Although oral rehydration therapy to correct and maintain the fluid and electrolyte balances during the acute diarrhoeal episode has substantially reduced the rate of acute complications, it has little effect on the course of acute diarrhoea. There is indirect evidence of disturbed intestinal microecology in children with rotavirus gastroenteritis and this in turn suggests that intestinal microbial balance may be linked with early cessation of rotavirus diarrhoea. Indeed, rotavirus infection appears to give rise to biphasic diarrhoea, the first phase being an osmotic diarrhoea and the second associated with overgrowth of specifically urease-producing bacteria (Isolauri *et al.*, 1994). These observations suggested that oral bacteriotherapy may be a promising means to counteract the disturbed microbial balance. Research on rotavirus gastroenteritis in infants is currently focusing on optimal diet. Recent works have highlighted that dietary supplementation with LAB counteracts the rise in urease activity (Isolauri *et al.*, 1994) and promotes the recovery of children with rotavirus diarrhoea through augmenting both local and systemic immune defence (Kaila *et al.*, 1992). Furthermore, specific IgA response has been endorsed, which is probably relevant in allowing protection against rotavirus reinfections (Kaila *et al.*, 1992).

The effectiveness of oral bacteriotherapy during rotavirus diarrhoea may be linked with the viability of LAB contained in the probiotic preparations and their ability to survive the gut transit, adhere to the mucosa and transiently colonize the gut. The adherent property significantly enhances the local immune response and stabilizes the mucosal barrier, resulting in reduced inadvertent transmission of antigens from

Table 6.2 Role of probiotics (lactic-acid producing bacteria (LAB)) in the treatment of human infections

- Rotavirus gastroenteritis
- Tropical acute watery diarrhoea
- *Clostridium difficile*-induced colitis
- Travellers' diarrhoea
- Bacterial vaginosis
- Urinary infections

the gut (Isolauri *et al.*, 1993). In contrast to the alleged importance of bacterial viability, immunostimulation and a beneficial clinical effect have also been shown using a supernatant obtained after centrifugation of LAB (Perdigon *et al.*, 1988). However, a recent study has demonstrated that, even though the clinical recovery from rotavirus infection was equal in patients receiving oral bacteriotherapy with viable or heat-inactivated LAB, viable bacteria stimulated specific IgA-antibody responses more effectively than inactivated bacteria (Kaila *et al.*, 1995).

There is also evidence that oral bacteriotherapy with LAB may accelerate recovery in cases of acute watery diarrhoea in children in developing countries, but no effect was observed in cases of bloody diarrhoea (Raza *et al.*, 1995).

In addition, daily doses of LAB are a well-accepted nutritional supplement by infants and mothers and result in an efficient colonization of the gastrointestinal tract. These observations further support the view that probiotics should be evaluated as an adjunct for diarrhoea control programmes at the community level as well as in hospital-based settings. Preliminary reports also indicate that LAB supplementation could be effective in improving the course of *C. difficile*-induced colitis (Rolfe, 1995). By contrast, there are conflicting reports about the effectiveness of oral bacteriotherapy in the treatment of travellers' diarrhoea (Oksanen *et al.*, 1990; Katelaris, Salam and Farthing, 1995).

In addition to their immunostimulating activity, LAB are also effective in the treatment of the diarrhoeal disorders listed above through other mechanisms. They produce lactic and acetic acids, hydrogen peroxide, protein inhibitors and antimicrobial substances, such as bacteriocin-like compounds, and this is crucial to the maintenance of colonization resistance and inhibition of the growth of pathogenic bacteria (De Simone *et al.*, 1995).

Several types of antibiotics have been used to treat bacterial vaginosis, but none was completely successful in either cure or prevention (Neri, Rabinerson and Kaplan, 1994). Several attempts have recently been made to treat bacterial vaginosis using physiological or natural substances containing LAB strains. This kind of treatment looks promising and may have a place in certain conditions, including pregnancy, in cases of recurrent inflammation, or as a prophylactic treatment before invasive gynaecological procedures or abdominal surgery in patients known to be affected (Neri, Rabinerson and Kaplan, 1994). It is still uncertain whether vaginal application of LAB reduces the infection rate in cystitis-prone women (Baerheim, Larsen and Digrares, 1994; Reid and Bruce, 1995).

A systemic immune response to LAB is frequently seen following their interaction with intestinal mucosa (Takahashi *et al.*, 1993). Whether this phenomenon could be detrimental to the therapeutic efficacy of administering LAB to patients with intestinal infections, such as rotavirus diarrhoea, remains to be established.

On the basis of currently available literature, no definite conclusions can be drawn. However, it is reasonable to conclude that probiotics are probably more helpful in the treatment of chronic diarrhoea because of the self-limiting nature of acute diarrhoea. In turn, the role of probiotics in preventing infectious diarrhoeas remains to be fully demonstrated.

At the other end of the spectrum of motility disorders of the gastrointestinal tract, some reports have shown that probiotic laxation may be effective in the treatment of constipation due to decreased motility, such as that commonly observed in the elderly (Jansen and Van der Waaij, 1995). Remarkably, this approach was not associated with such side-effects as dehydration and disturbances in the electrolyte balance frequently seen in subjects receiving traditional laxatives.

6.16   PATHOGENIC POTENTIAL OF PROBIOTICS

LAB are commonly considered to be commensal or beneficial participants in human microbial ecology, but they are also associated with some human diseases (dental caries, rheumatic valvular diseases, pneumonia, sepsis and infective endocarditis) (Davies, James and Hawkey, 1986; Chomarat and Espinouse, 1991; Harty *et al.*, 1994; Jones *et al.*, 1994; Ozaki *et al.*, 1994) However, there is no evidence of them acting as primary pathogens in any of these disorders. LAB have recently been identified as potential emerging pathogens in elderly and immunocompromised patients, particularly those receiving broad-spectrum antibiotic therapy (Chomarat and Espinouse, 1991; Maskell and Pead, 1992). The identification of potential pathogenic traits amongst LAB will therefore improve the selection of these micro-organisms for probiotic purposes.

Interestingly, the identification of infective endocarditis isolates as primarily belonging to two species, namely *L. rhamnosus* and *L. paracasei*, suggests that they may have a greater pathogenic potential than other species, even though these findings may be in part a reflection of their relative numbers in the oral cavity (Harty *et al.*, 1994).

The ability to aggregate human platelets by LAB is considered to be a possible pathogenic trait in the progression of infective endocarditis (Harty *et al.*, 1994). In addition, LAB involved in infective endocarditis have the potential to bind both fibronectin and fibrinogen as well as the proteins that make up the extracellular matrix of endothelial cells (Harty *et al.*, 1994). Available data also indicate that the LAB species most frequently isolated from cases of endocarditis, *L. rhamnosus* and *L. paracasei*, produce enzymes such as glycosidases and arylamidases that would enable the breakdown of human glycoproteins and the synthesis and lysis of human fibrin clots (Oakey *et al.*, 1995). These characteristics are considered to aid the colonization and survival of bacterial infecting an endocardial vegetation.

The possibility that some LAB species could be potentially pathogenic, albeit opportunistic pathogens with a low infection rate, should be borne in mind in view of their widespread use as dietary supplements.

6.17 REFERENCES

Aattouri, N. and Lemonnier, D. (1995) Involvement of CD lymphocytes in the production of interferon induced by non-pathogenic bacteria, in *Probiotics: Prospects of Use in Opportunistic Infections* (eds R. Fuller, P.J. Heidt, V. Rusch and D. Van der Waaij), Institute for Microbiology and Biochemistry, Old Herborn University, Herborn-Dill, pp. 81–8.

Abraham, R. and Ogra, P.L. (1994) Mucosal microenvironment and mucosal response. *Am. J. Trop. Med. Hyg.*, **50**, 3–9.

Akeda, H., Mitsuyama, M., Tatsukawa, K. *et al.* (1981) The synergistic contribution of macrophages and antibody to protection against *Salmonella typhimurium* during early phase of infection. *J. Gen. Microbiol.*, **123**, 209–13.

Antonaci, S. and Jirillo, E. (1985) Relationship between immune system and Gram-negative bacteria: monocyte chemotaxis induced by supernatants from human peripheral blood OKT8+ lymphocytes stimulated with smooth and rough *Salmonella* strains. *Cell. Immunol.*, **95**, 258–61.

Apperloo-Renkema, H.Z., Bootsma, H., Mulder, B.I. *et al.* (1994) Host-microflora interaction in systemic lupus erythematosus (SLE): colonization resistance of the indigenous bacteria of the intestinal tract. *Epidemiol. Infect.*, **112**, 367–73.

Apperloo-Renkema, H.Z., Bootsma, H., Mulder, B.I. *et al.* (1995) Host–microflora interaction in systemic lupus erythematosus (SLE): circulating antibodies to the indigenous bacteria of the intestinal tract. *Epidemiol. Infect.*, **114**, 133–41.

Baerheim, A., Larsen, E. and Digranes, A. (1994) Vaginal application of lactobacilli in the prophylaxis of recurrent lower urinary tract infection in women. *Scand. J. Prim. Health Care*, **12**, 239–43.

Balmer, S.E. and Wharton, B.A. (1989) Diet and faecal flora in the newborn: breast milk and infant formula. *Arch. Dis. Child.* **4**, 1672–7.

Befus, A.D. (1993) Neuromodulation of gastrointestinal immune and inflammaory responses, in *Immunopharmacology of the Gastrointestinal System* (ed. J.L. Wallace), Academic Press, New York, pp. 1–14.

Bland, P.W. (1987) Antigen presentation by gut epithelial cells: secretion by rat enterocytes of a factor with IL-1-like activity. *Adv. Exp. Med. Biol.*, **216**, 219–25.

Butler, L.D., Layman, N.K., Cain, R.I., *et al.* (1988) Interleukin 1-induced pathophysiology: induction of cytokines, development of histopathologic changes and immunopharmacologic intervention. *Clin. Immunol. Immunopathol.*, **53**, 400–21.

Cerf-Bensussan, N. and Guy-Grand, D. (1991) Intestinal intraepithelial lymphocytes. *Gastroenterol. Clin. N. Amer.*, **20**, 549–76.

Chen, W., Harp, J.A. and Harmsen, A.G. (1993a) Requirements for CD4+ cells and gamma interferon in resolution of established *Cryptosporidium parvum* infection in mice. *Infect. Immun.*, **61**, 3928–32.

Chen, W., Harp, J.A. Harmsen, A.G. and Havell, E.A. (1993b) Gamma interferon functions in resistance to *Cryptosporidium parvum* infection in severe combined immunodeficient mice. *Infect. Immun.*, **61**, 3548–51.

Chomarat, M. and Espinouse, D. (1991) *Lactobacillus rhamnosus* septicemia in patients with prolonged aplasia receiving ceftazidime–vancomycin. *Eur. J. Clin. Microbiol. Infect. Dis.*, **10**, 44.

Collins, S.M. (1993) The immune modulation of intestinal motor function, in *Immunopharmacology of the Gastrointestinal System* (ed. J.L. Wallace), Academic Press, New York, pp. 41–50.

Davidson, P.M. and Hoover, D.G. (1993) Antimicrobial components from lactic acid bacteria, in *Lactic Acid Bacteria* (eds S. Salminen and A. von Wright), Marcel Dekker, New York, pp. 127–59.

Davies, A.J., James, P.A. and Hawkey, P.M. (1986) *Lactobacillus* endocarditis. *J. Infect.*, **12**, 169–74.

Deem, R.L., Shanahan, F. and Targan, S.R. (1991) Triggered human T cells release tumour necrosis factor-alpha and interferon-gamma which kill human colonic epithelial cells. *Clin. Exp. Immunol.*, **83**, 79–84.

De Simone, C., Famularo, G., Harp, J.A. *et al.* (1995) Effect of lactobacilli on *Cryptosporidium parvum* infection in man and animals. *Microecol. Ther.*, **25**, 32–6.

De Simone, C., Bianchi Salvadori, B., Negri, R. *et al.* (1986) The adjuvant effect of yogurt on production of gamma-interferon by Con A stimulated human peripheral blood lymphocytes. *Nutr. Reports Int.*, **3**, 419–31.

De Simone, C., Grassi, P.P., Bianchi Salvadori, B. *et al.* (1988a) Adherence of specific yogurt microorganisms to human peripheral blood lymphocytes. *Microbios*, **55**, 49–57.

De Simone, C., Jirillo, E., Bianchi Salvadori, B. *et al.* (1988b) Stimulation of host resistance by a diet supplemented with yogurt, in *Immunomodulators and Nonspecific Host Defence Mechanisms against Microbial Infections* (eds K.N. Masihi and W. Lange), Pergamon Press, Oxford, pp. 279–88.

De Simone, C., Tzantzoglou, S., Baldinelli, L. *et al.* (1988c) Enhancement of host resistance against *Salmonella typhimurium* infection by a diet supplemented with yogurt. *Immunopharmacol. Immunotoxicol.*, **10**, 399–408.

De Simone, C., Vesely, R., Bianchi Salvadori, B. and Jirillo, E. (1993) The role of probiotics in modulation of the immune system in man and in animals. *Int. J. Immunother.*, **9**, 23–8.

Dichelte, G., Kaspereit, F. and Sedlacek, H.H. (1984) Stimulation of cell-mediated immunity by Bestatin correlates with reduction of bacterial persistence in experimental chronic *Salmonella typhimurium* infection. *Infect. Immun.*, **44**, 168–72.

Dumonceau, J.M., Adler, M., Cremer, M. and Van Gossum, A. (1994) The role of infectious agents in Crohn's disease. *Acta Gastroenterol. Belgica*, **57**, 301–5.

Edwards, C. (1994) The nutritional impact of the intestinal microflora. *SAMY*, **32**, 38–9.

Englyst, H.N., Kingman, S.M. and Cummings, J.H. (1992) Classification of measurement of nutritionally important starch fractions. *Eur. J. Clin. Nutr.*, **46**, S33–S51.

Fayer, R., Speer, C.A. and Dubey, J.P. (1990) General biology of *Cryptosporidium*, in *Cryptosporidiosis of Man and Animals* (eds J.P. Dubey, C.A. Speer and R. Fayer), CRC Press, Boca Raton, FL, pp. 1–29.

Ferreira, R., Forsyth, L.E., Richman, P.I. *et al.*, (1990) Changes in the rate of crypt epithelial cell proliferation and mucosal morphology induced by a T-cell-mediated response in human small intestine. *Gastroenterology*, **98**, 1255–63.

Freter, R. and Nader de Macias, M.E. (1995) *Factors Affecting the Colonisation of the Gut by Lactobacilli and Other Bacteria* (eds R. Fuller, P.J. Heidt, V. Rusch and D. Van der Waaij), Institute for Microbiology and Biochemistry, Old Herborn University, Herborn Dill, pp. 19–34.

Fuller, R. (1995) Probiotics: their development and use, in *Probiotics: Prospects of Use in Opportunistic Infections* (eds R. Fuller, P.J. Heidt, V. Rusch and D. Van der Waaij), Institute for Microbiology and Biochemistry, Old Herborn University, Herborn Dill, pp. 1–8.

Fujihashi, K., Taguchi, T., McGhee J.R. *et al.* (1990) Regulatory function for murine intraepithelial lymphocytes: two subsets of CD3+, T-cell receptor-1+ intraepithelial lymphocyte T cells abrogate oral tolerance. *J. Immunol.*, **145**, 2010–19.

Gilliland, S.E. (1981) Enumeration and identification of lactobacilli in feed supplements marketed as sources of *Lactobacillus acidophilus*. *Okla. Agric. Exp. Sta.*, **108**, 61–3.

Gilliland, S.E. and Speck, L.M. (1977) Antagonistic action of *Lactobacillus acidophilus* toward intestinal and foodborne pathogens in associative cultures. *J. Food Prod.*, **49**, 820–3.

Gundel, R.H. and Letts, G.L. (1994) Adhesion molecules and the modulation of mucosal inflammation, in *Immunopharmacology of Epithelial Barriers* (ed. R. Goldie), Academic Press, New York, pp. 71–84.

Harp, J.A., Chen, W. and Harmsen, A.G. (1992) Resistance of severe combined immunodeficient mice to infection with *Cryptosporidium parvum*: the importance of intestinal microflora. *Infect. Immun.*, **60**, 3509–12.

Harp, J.A., Woodmansee, D.B. and Moon, H.W. (1990) Resistance of calves to *Cryptosporidium parvum*: effects of age and previous exposure. *Infect. Immun.*, **58**, 2237–40.

Harty, D.W.S., Oakey, H.J., Patrikakis, M. *et al.*, (1994) Pathogenic potential of lactobacilli. *Int. J. Food Microbiol.*, **24**, 179–89.

Hatcher, G.E. and Lambrecht, R.S. (1993) Augmentation of macrophage phagocytic activity by cell-free extracts of selected lactic acid-producing bacteria. *J. Dairy Sci.*, **76**, 2485–9.

Hitchins, A.D., Wells, P., McDonough, F.E. and Wong, N.P. (1985) Amelioration of the adverse effects of a gastrointestinal challenge with *Salmonella enteridis* on weaning rats by a yogurt diet. *Am. J. Clin. Nutr.*, **41**, 92–100.

Holmgren, J., Frylung, J. and Larsson, H. (1989) Gamma-interferon-mediated down-regulation of electrolyte secretion by intestinal epithelial cell: a local immune mechanism? *Scand. J. Immunol.*, **30**, 499–503.

Howie, P.W., Forsyth, J.S., Ogston, S.A. *et al.*, (1990) Protective effect of breast feeding against infection. *BMJ*, **300**, 11–16.

Husband, A. and Dunkley, M. (1990) Helper T cell control of mucosal immune responses. *Today's Life Sci.*, **2**, 22–31.

Isolauri, E., Kaila, M., Arvola, T. *et al.* (1993) Diet during rotavirus enteritis affects jejunal permeability to macromolecules in suckling rats. *Pediatr. Res.*, **33**, 548–53.

Isolauri, E., Kaila, M., Mykkanen, H. *et al.* (1994) Oral bacteriotherapy for viral gastroenteritis. *Dig. Dis. Sci.*, **39**, 2595–600.

Jansen, G.J. and Van der Waaij, D. (1995) Prospects of use in opportunistic infections: review of the internal discussion, in *Probiotics: Prospects of Use in Opportunistic Infections* (eds R. Fuller, P.J. Heidt, V. Rusch and D. Van der Waaij), Institute for Microbiology and Biochemistry, Old Herborn University, Herborn Dill, pp. 173–84.

Jones, S.D., Fullerton, D.A., Zamora, M.R. *et al.* (1994) Transmission of *Lactobacillus* pneumonia by a transplanted lung. *Ann. Thorac. Surg.*, **58**, 887–9.

Kagnoff, M.F. (1996) Mucosal immunology: new frontiers. *Immunol. Today*, 57–9.

Kaila, M., Isolauri, E., Soppi, E. *et al.* (1992) Enhancement of the circulating antibody secreting cell response in human diarrhea by a human lactobacillus strain. *Pediatr. Res.*, **32**, 141–4.

Kaila, M., Isolauri, E., Saxelin, M. *et al.* (1995) Viable versus inactivated *Lactobacillus* strain GG in acute rotavirus diarrhoea. *Arch. Dis. Child.*, **72**, 51–53.

Kanerud, L., Scheynius, A., Nord, C.E. and Hafstrom, I. (1994) Effect of sulphasalazine on gastrointestinal microflora and on mucosal heat shock protein expression in patients with rheumatoid arthritis. *Br. J. Rheumatol.*, **33**, 1039–48.

Katelaris, P.H., Salam, I. and Farthing, M.J.G. (1995) Lactobacilli to prevent traveler's diarrhea? *N. Engl. J. Med.*, **333**, 1360–1.

Keshav, S., Lawson, L., Chung, L.P. *et al.* (1990) Tumour necrosis factor mRNA localized to Paneth cells of normal murine intestine by *in situ* hybridisation. *J. Exp. Med.*, **171**, 327–32.

Kitazawa, H., Matsumura, K., Itoh, T. and Yamaguchi, T. (1992) Interferon induction in murine peritoneal macrophage by stimulation with *Lactobacillus acidophilus*. *Microbiol. Immunol.*, **36**, 311–15.

Kitazawa, H., Tomioka, Y., Matsumura, K. *et al.* (1994) Expression of mRNA encoding IFN alpha in macrophages stimulated with *Lactobacillus gasseri*. *FEMS Microbiol Lett.*, **120**, 315–21.

Kvale, D., Lovhaug, D., Sollid, L.M. and Brandtzaeg, P. (1988) Tumour necrosis factor-alpha upregulates expression of secretory component, the epithelial receptor for polymeric Ig. *J. Immunol.*, **140**, 3086–9.

Link-Amster, H., Rochat, F., Saudan, K.Y. *et al.* (1994) Modulation of a specific humoral immune response and changes in intestinal flora mediated through fermented milk. *FEMS Immunol. Med. Microbiol.*, **10**, 55–63.

London, S.D., Rubin, D.H. and Cebra, J.J. (1987) Gut mucosal immunization with reovirus serotype 1/L stimulates virus specific cytotoxic T-cell precursors as well as IgA memory cells in Peyer's patches. *J. Exp. Med.*, **165**, 830–7.

Maskell, R. and Pead, L. (1992) 4-Fluorochinolones and *Lactobacillus* spp. as emerging pathogens. *Lancet*, **339**, 929.

Mayrhofer, G. (1994) Epithelial disposition of antigen, in *Immunopharmacology of Epithelial Barriers* (ed. R. Goldie), Academic Press, New York, pp. 19–70.

McGhee, J.R., Mestecky, J., Elson, C.O. and Kiyono, H. (1989) Regulation of IgA synthesis and immune response by T cells and interleukins. *J. Clin. Immunol.*, **9**, 175–99.

Motyka, B. and Reynolds, J.R. (1991) Apoptosis is associated with the extensive B cell death in the sheep ileal Peyer's patch and the chicken bursa of Fabricius: a possible role in B cell selection. *Eur. J. Immunol.*, **21**, 1951–8.

Muscettola, M., Massai, L., Tanganelli, C. and Grasso, G. (1994) Effects of lactobacilli on interferon production in young and aged mice. *Ann. NY Acad. Sci.*, **717**, 226–32.

Neri, A., Rabinerson, D. and Kaplan, B. (1994) Bacterial vaginosis: drugs versus alternative treatment. *Gynecol. Obstet. Survey*, **49**, 809–13.

Nicaise, P., Gleizes, A., Forestier, F. *et al.* (1993) Influence of intestinal bacterial flora on cytokine (IL-1, IL-6 and TNF-alpha) production by mouse peritoneal macrophages. *Eur. Cyt. Network*, **4**, 133–8.

Oakey, H.J., Harty, D.W.S. and Knox, K.W. (1995) Enzyme production by lactobacilli and the potential link with infective endocarditis. *J. Appl. Bacteriol.*, **78**, 142–8.

Oksanen, P.J., Salminen, S., Saxelin, M. *et al.* (1990) Prevention of travellers' diarrhoea by *Lactobacillus* GG. *Ann. Med.*, **22**, 53–6.

Osada, K., Nagira, K., Teruya, K. *et al.* (1994) Enhancement of interferonbeta production with sphingomyelin from fermented milk. *Biother.*, **7**, 115–23.

Ozaki, K., Matsuo, T., Nakae, H. *et al.* (1994) A quantitative comparison of selected bacteria in human carious dentine by microscopic counts. *Caries Res.*, **28**, 137–45.

Perdigon, G., Nader de Macias, M.E., Alvarez, S., Oliver, G., de Ruiz-Holgado, A.P. (1988) Systemic augmentation of the immune response in mice by feeding fermented milks with *Lactobacillus casei* and *Lactobacillus acidophilus*. *Immunology*, **63**, 17–23.

Perdue, M.H. and McKay, D.M. (1993) Immunomodulation of the gastrointestinal epithelium, in *Immunopharmacology of the Gastrointestinal System* (ed. J.L. Wallace), Academic Press, New York, pp. 15–39.

Pereyra, B.S. and Lemonnier, D. (1993) Induction of human cytokines by bacteria used in dairy foods. *Nutr. Res.*, **13**, 1127–40.

Popova, P., Guencheva, G., Davidkova, G. *et al.* (1993) Stimulating effect of DEODAN (an oral preparation from Lactobacillus bulgaricus 'LB51') on monocytes/macrophages and host resistance to experimental infections. *Int. J. Immunopharmac.*, **15**, 25–37.

Raza, S., Graham, S.M., Allen, S.J. *et al.* (1995) *Lactobacillus* GG promotes recovery from acute nonbloody diarrhea in Pakistan. *Pediatr. Infect. Dis. J.*, **14**, 107–11.

Reid, G. and Bruce, A.W. (1995) Low vaginal pH and urinary tract infection. *Lancet*, **346**, 1704.

Rolfe, R.D. (1995) Probiotics: prospects for use in *Clostridium difficile*-associated intestinal disease, in *Probiotics: Prospects of Use in Opportunistic Infections* (eds R. Fuller, P.J. Heidt, V. Rusch and D. Van der Waaij), Institute for Microbiology and Biochemistry, Old Herborn University, Herborn Dill, pp. 47–66.

Rosenberg, S.A., Lotze, M.T. and Mule, J.J. (1988) New approaches to immunotherapy of cancer using interleukin-2. *Ann. Intern. Med.*, **108**, 853–64.

Rutgeerts, P., Geboes, K., Peerers, M. *et al.* (1991) Effect of fecal stream diversion on recurrence of Crohn's disease in the neoterminal ileum. *Lancet*, **338**, 771–4.

Santos, L.M.B., Lider, O., Audette, J. *et al.* (1990) Characterization of immunomodulatory properties and accessory cell function of small intestinal epithelial cells. *Cell. Immunol.*, **127**, 26–34.

Schiffrin, E.J., Rochat, F., Link-Amster, H. *et al.* (1995) Immunomodulation of human blood cells following the ingestion of lactic acid bacteria. *J. Dairy Sci.*, **78**, 491–7.

Shinebaum, R., Neumann, V.C., Cooke, E.M. and Wright, V. (1987) Comparison of faecal florae in patients with rheumatoid arthritis and controls. *Br. J. Rheumatol.*, **26**, 329–33.

Shirota, K., LeDug, L., Yuan, S. and Jothy, S. (1990) Interleukin-6 and its receptor are expressed in human intestinal epithelial cells. *Virchows Arch. Cell. Pathol.*, **58**, 303–8.

Smeyanov, V.V., Maltseva, N.N., Bossart, W. and Korshunov, V.M. (1992) Influence of *Lactobacillus acidophilus* Solco on immunological parameters of totally decontaminated mice under the conditions of genera germ-free isolation. *Zhurn. Mikrobiol. Epidemiol. Immunobiol.*, **11**, 12–15.

Struthers, G.R. (1986) *Clostridium perfringens* and rheumatoid arthritis. *Br. J. Rheumatol.*, **25**, 419–20.

Takahashi, T., Oka, T., Iwana, H. *et al.* (1993) Immune response of mice to orally administered lactic acid bacteria. *Biosci. Biotechnol. Biochem.*, **57**, 1557–60.

Terada, K., Okuhara, E. and Kawarada, Y. (1991) Antigen DNA isolated from immune complexes in plasma of patients with systemic lupus erythematosus hybridizes with the *Escherichia coli* lac Z gene. *Clin. Exp. Immunol.*, **85**, 66–9.

Thompson, S.J. and Elson, C.J. (1993) Susceptibility to pristane-induced arthritis is altered with changes in bowel flora. *Immunol. Lett.*, **36**, 227–32.

Tomita, S., Myones, B.L. and Shulman, S.T. (1993) *In vitro* correlates of the *L. casei* animal model of Kawasaki disease. *J. Rheumatol.*, **20**, 362–7.

Tsuneoka, K., Ishihara, H., Dimchev, A.B. *et al.* (1994) Timing in administration of a heat-killed *Lactobacillus casei* preparation for radioprotection in mice. *J. Radiat. Res.*, **35**, 147–56.

Umesaki, Y., Setoyama, H., Matsumoto, S. and Okada, Y. (1993) Expansion of alpha-beta T-cell receptor-bearing intestinal intraepithelial lymphocytes after microbial colonization in germ-free mice and its independence from thymus. *Immunology*, **79**, 32–7.

Ungar, B.L.P. (1990) Cryptosporidiosis in humans (*Homo sapiens*), in *Cryptosporidiosis of Man and Animals* (eds J.P. Dubey, C.A. Speer and R. Fayer), CRC Press, Boca Raton, FL, pp. 59–82.

Ungar, B.L.P., Kao, T.-C., Burris, J.A. and Finkelman, F.D. (1991) *Cryptosporidium* infection in an adult mouse model. Independent roles for IFN-gamma and CD4+ T lymphocytes in protective immunity. *J. Immunol.*, **147**, 1014–22.

Wellmann, W., Fink, P.C., Benner, F. and Schmidt, F.W. (1986) Endotoxaemia in active Crohn's disease. Treatment with whole gut irrigation and 5-aminosalicylic acid. *Gut*, **27**, 814–20.

Zu, S.-X., Fang, G.-D., Fayer, R, and Guerrant, R.L. (1992) Cryptosporidiosis: pathogenesis and immunology. *Parasitol. Today*, **8**, 24–7.

CHAPTER 7

Probiotics in cattle

J.T. Huber

7.1 INTRODUCTION

A trademark of the research reported from studies where probiotics or direct-fed microbials have been included in cattle rations has been the variability of observed results. Such variability has not only been shown in controlled experiments, but also in commercial settings. This author has known of a number of producers who have used one or more of the probiotics for a period with favourable results and have withdrawn or switched to another probiotic because they sensed that the initial additive(s) became ineffective. A possible reason for this variability in response is an adaptation to the probiotic by a particular environment of micro-organisms or a shift in feeding or management conditions which might have rendered a fungal or bacterial culture ineffective.

Despite a lack of consistency of effects often shown from feeding probiotics, their acceptance and use in cattle production systems have been steadily increasing over the past decade. Certainly, they are one of the least expensive additives available with daily costs per animal for fungal cultures ranging from $US0.03 to $US0.07. Bacterial cultures tend to be more expensive. In addition to animal production benefits such as increased milk yields and weight gains (which overall have been positive, but variable), subtle effects such as less days off feed, a lower incidence of digestive upsets, better body condition of animals, less days in sick pen, and others, have been attributed to feeding of the various probiotics.

This chapter reviews much of the recent probiotic research conducted with baby calves, beef cattle and lactating cows. Our goal is to evaluate objectively the effectiveness and relative benefits of supplementing

Probiotics 2: Applications and practical aspects.
Edited by R. Fuller.
Published in 1997 by Chapman & Hall. ISBN 0 412 73610 1

cattle rations with probiotics. The chapter concludes with a brief discussion on market penetration and proposed mechanism(s) of action.

7.2 BABY CALVES

There have been a number studies reported on feeding probiotics to young calves, and the practice is accepted in many areas. Feeding of yeast to calves was reported as early as 1924 (Eckles *et al.*, 1924), and numerous subsequent reports have suggested beneficial effects of probiotic addition early in the life of the calf (Table 7.1). Several, but not all, of these studies are highlighted in this chapter in order to reach a conclusion on effectiveness of feeding probiotics to baby calves.

Table 7.1 Effects of probiotic feeding to baby calves

Probiotic	Vehicle	Observed results	References
YC[2]	Starter	Greater feed intake/ weight gain Stabilized fermentation	Fallon and Harte, 1987; Hughes, 1988
Brewers yeast	Starter	Less fever and antibiotic use	Seymour *et al.*, 1995
AO	Starter	Ate starter faster, weaned earlier, accelerated rumen fermentation	Beharka *et al.*, 1990
	Starter	Improved feed intake/ weight gains	Allison and McGraw, 1989
	Diet of dam	Improved weight gains in suckling calves	Wiedmeier, 1989
Lactobacillus spp.	Milk replacer	Decreased faecal coliforms	Ellinger *et al.*, 1978; Bruce *et al.*, 1979; Gilliland *et al.*, 1980
	Milk replacer	Reduced incidence of diarrhoea	Bechman *et al.*, 1977; Beeman, 1985; Bonaldi *et al.*, 1986
	Milk replacer	Improved feed intake/ weight gain	Wren, 1987; Lee and Botts, 1988; Higginbotham *et al.*, 1993
	Milk replacer	Replaced antibiotic	Morrill *et al.*, 1995
Streptococcus faecium	Milk replacer	Increased feed intake/ decreased scouring	Hefel, 1980; Ozawa *et al.*, 1983; Maeng *et al.*, 1987; Tournut, 1989: Szovil *et al.*, 1987

YC[2], YeaSacc
AO, Amaferm

7.2.1 Supplementation to liquid diets

Lactobacillus-type cultures were added to the liquid diets of newborn dairy calves in 10 studies. Three studies (Ellinger, Muller and Glantz, 1978; Bruce *et al.*, 1979; Gilliland *et al.*, 1980) reported decreased counts of faecal coliform bacteria, three detected reduced incidence of diarrhoea (Bechman, Chambers and Cunningham, 1977; Beeman, 1985; Bonaldi *et al.*, 1986), and higher feed intakes and weight gains were shown by two (Wren, 1987; Lee and Botts, 1988). More recently, Higginbotham and Bath (1993) reported that although there were no significant differences in average daily gains of calves fed both viable and non-viable microbial products, they tended to have higher weight gains during the early weeks of the feeding period. Starter intake was higher for the calves fed the non-viable product during the fifth week of treatment, and no differences among treatments were noted for faecal counts of coliform, streptococci or lactobacilli bacteria. Although Morrill *et al.* (1995) did not show any improvement in growth performance from adding a lactobacilli-type antibiotic to a milk replacer containing all milk protein, when compared to a similar replacer containing antibiotics, they suggested that calves had benefited as much by the probiotic as they had by the antibiotics added to the control milk replacer. When *Enterococcus faecium* was added to the milk replacer of calves in five studies (Hefel, 1980; Ozawa *et al.*, 1983; Maeng, Kim and Shin, 1987; Svozil *et al.*, 1987; Tournut, 1989), increased starter intake and decreased scouring were reported.

7.2.2 Supplementation to starter diets

Essentially, all studies where probiotics were added to starter for baby calves used fungal cultures which are described in the footnote below. Addition of 2 g kg^{-1} of YC[2]* to the starter diet for young calves weaned at 5–6 weeks in two experiments resulted in greater feed intakes and a 25% improvement in body weight gain (BWG) ($P < 0.05$) to weaning (Fallon and Harte, 1987; Hughes, 1988). During postweaning, up to 12 weeks when treatment ended, the same trends in intakes and gains continued, but effects of YC[2] were not significant. In the study of Fallon and Hart (1987), corn gluten replaced the barley in the starter diet. The added YC[2] apparently stabilized fermentation of the developing rumen, which led the authors to conclude that the YC[2] allowed calves on the gluten diet to maintain feed intakes equal to those fed barley plus YC[2].

---

*YC[1], Diamond V yeast culture (YC) produced by Diamond V Mills, Cedar Rapids, IA; YC[2], YeaSacc, produced by Alltech Biotechnology, Nicholasville, KY. AO, Amaferm, a viable *Aspergillus oryzae* culture produced by Biozyme, Inc., St. Joseph, MO. Both YC[1] and YC[2] are viable cultures of *Saccharomyces cerevisiae*.

Addition of 1% brewers yeast (0.5 g day^{-1}) to starter fed dairy calves during the preweaning period (1–46 days) reduced the incidence of fever and the number of antibiotic treatments, but had no visible effect during the 43-day postweaning period (Seymour, Nocek and Siciliano Jones, 1995). Hence, yeast addition to the dry feed apparently reduced susceptibility of calves to infections prior to weaning, but had no effect on weight gains or feed efficiency during preweaning or postweaning.

Kansas workers (Beharka, Nagaraja and Morrill, 1990) tested effects of supplementation of the starter diet with *Aspergillus oryzae* (AO) extract on performance and development of rumen function in baby calves weaned when they consumed 550 g day^{-1} of starter. Calves fed AO (0.5–3 g day^{-1}) ate starter faster and were ready to wean 1 week earlier than unsupplemented calves. This was attributed to accelerated rumen development in calves fed AO compared with control calves. Greater concentrations of total VFA, acetate and propionate, and higher counts of total anaerobes, hemicellulolytic and pectinolytic bacteria, with a tendency towards higher cellulolytic counts were also found in the AO supplemented calves.

An indirect effect on calves from feeding the AO culture to their dams was shown by Wiedmeier (1989). Milk yields of the cows were apparently increased by AO supplementation because ADG of calves from cows receiving AO were 0.80 kg day^{-1} compared with 0.57 kg day^{-1} for controls. Weight gains of cows were not affected by AO supplementation.

In conclusion effects of supplementing milk replacers with *Lactobacillus* and *Streptococcus* cultures include decreased coliform counts, reduced scouring and improved growth performance. Although addition of fungal probiotics of calf starters did not usually show dramatic improvements in performance, most studies have demonstrated benefits such as earlier intake of starter, greater stability of feed consumption patterns, growth increases or faster development of rumen fermentative action. Specific situations will dictate which of a large array of available probiotics might benefit calf health and performance, but many of the studies have shown favourable results during this critical and difficult period of the animal's life.

7.3 BEEF CATTLE

Effect on performance, rumen fermentation, nutrient utilization and various aspects of metabolism have been considered in evaluating fungal probiotics for growing, non-lactating beef cattle. Although performance has not been dramatically improved by supplementation with either YC[1], YC[2] or AO, beneficial effects have usually been reported. Adams *et al.* (1981) showed increased intakes ($P < 0.05$) in steers fed 1.85% YC^1 in a 50:50 concentrate:forage diet. Even though ADG and feed efficiencies were numerically higher with YC^1-supplemented steers, differences were not significant. A similar response of higher feed intakes (5.6 vs. 5.3 kg day^{-1}; $P < 0.05$) resulted from supplementing YC^2, but only a tendency towards increased gains and feed efficiencies was observed with beef bulls fed a barley:soybean meal diet from 135 to 450 kg liveweight (Edwards *et al.*, 1990). Canadian workers (Mir and Mir, 1994) fed three diets sequentially to Hereford steers treated with 0 or 10 g day^{-1} live yeast culture (YC^2) in two separate years. The three diets were: 75% alfalfa silage and 25% barley, 96% corn silage and 4% soybean meal, or 75% dry rolled barley and 25% alfalfa hay. Steers were re-randomized each time diets were switched. Increased feed efficiency was observed for steers fed YC^2 during the first year, but there were no differences between treatments in other performance or carcass characteristics during either of the years.

Parra and DeConstanzo (1992) compared effects of feeding YC^2 for the initial 28 days after cattle were moved to a feedlot or for the entire 58-day feeding period. Greater weight gains (1.48 vs. 1.23 kg day^{-1}) and feed efficiencies (7.73 vs. 8.80 kg dry matter (DM) kg^{-1} gain) were noted from 32 to 58 days in steers fed YC^2 compared with controls. The improved performance was attributed to higher DM digestibility for steers fed YC^2 (84.3 vs 76.1%). Montana workers (Hudyma and Gray, 1993) reported significantly higher weight gains (0.82 vs. 0.69 kg day^{-1}), feed intakes (6.66 vs. 6.38 kg day^{-1}) and feed conversion (8.14 vs. 9.28 kg DM kg^{-1} gain) from addition of YC^1 to a corn silage diet for crossbred steers.

One factor that may influence response of feedlot cattle to probiotics is the type of diet to which they are added. From the review of Wallace and Newbold (1992), ADG of cattle from nine comparisons where more than 50% concentrate was fed were increased only 3.8% by YC addition (1.38 vs. 1.33 kg day^{-1}); whereas, ADG in seven comparisons using high forage diets were 10.6% higher for cattle receiving YC (1.04 vs. 0.93 kg day^{-1}).

North Dakota workers (Olson *et al.*, 1994a) studied the effect of YC^2 supplementation in ruminally cannulated beef steers and heifers grazing a mixed-grass prairie pasture from late June to early November. Dietary composition, intake, and *in situ* nutrient disappearance were determined. The data showed that forage samples from steers fed YC^2 had greater soluble N and *in vitro* DM disappearance than those from control steers. Intakes of DM were increased by YC^2 during the early season and *in situ* neutral detergent fibre (NDF) and crude protein (CP) degradabilities were higher for YC^2 steers at various times during the grazing season. In a companion paper from the same series of studies, Olson *et al.* (1994b) demonstrated variable effects of YC^2 on rumen fermentation. However, steers receiving YC^2 in late July had greater ($P < 0.07$) duodenal flow of bacterial N, which the authors attributed to increased organic matter (OM) digestibility.

Age of feeder cattle also has been suggested as a factor influencing response to YC cultures. Again, data from three comparisons were summarized from the review of Wallace and Newbold (1992), which showed only slightly larger increases in weight gains (8.3%) for smaller cattle (144 kg) on YC, compared with increases of 6.3% in larger animals (334 kg) fed the additive. An example of small cattle responding to yeast was early weaned beef calves fed a high energy diet for 55 days in the study by Anderson (1993), who reported that supplementation of YC^2 ($10 \, g \, head^{-1} day^{-1}$) increased average daily gain (1.08 vs. $0.97 \, kg \, day^{-1}$; $P < 0.05$), but did not significantly affect feed intake (5.99 vs. $5.88 \, kg \, day^{-1}$) or feed conversion (5.56 vs. $6.04 \, kg$ DM kg^{-1} gain).

Cole, Purdy and Hutcheson (1992) reported the results of two experiments conducted to determine the influence of YC^2 on health, performance and response to a challenge with infectious bovine rhinotracheitis virus (IBRV) in beef feeder calves. Results showed no significant effect of yeast culture on health and performance of calves in either study. However, morbid calves fed YC^2 had fewer days on antibiotic therapy than controls. Moreover, calves challenged intranasally with IBRV which received YC^2 maintained heavier weights and higher DM intakes than those which did not. In the same article, lambs fed YC^2 were higher in retained N and tended to have greater retention of Zn and Fe than control lambs. The authors suggested that the beneficial effects of YC^2 on health of ruminants might be mediated through N, Fe and/or Zn metabolism.

7.3.1 *Aspergillus* cultures

Cultures of AO have been tested with steers fed high forage diets under grazing and feedlot conditions. Coffey *et al.* (1990) reported that AO did not affect gain, intake or feed conversion of steers offered fungus-infected tall fescue or bromegrass screenings. Moreover, the AO did not alleviate the classical symptoms associated with consumption of the infected fescue such as reduced feed intake, increased rectal temperatures and reduced serum prolactin. In another study, the same group (Coffey *et al.*, 1993) compared fungus-infected (FI) and fungus-free fescue (FF) hay with and without AO in the supplement. The ruminally cannulated beef heifers tended to consume more feed when offered AO and FF fescue. However, again the negative effect of FI fescue was not overcome with AO and nor did AO affect *in situ* or *in vitro* digestibilities or digestion rate of feed. Similarly, Kentucky workers (Ely *et al.*, 1990) showed no effect on weight gains or body temperatures in steers grazing tall fescue pasture when AO was added to a trace-mineralized salt mixture. They suggested that the salt intake was too variable to use it as a vehicle for AO feeding.

Contrary to the previous studies with fescue, Caton *et al.* (1993) demonstrated that AO supplementation of steers grazing bromegrass pasture increased forage OM intake and *in vitro* DM digestibility. Ruminal and total tract digestibilities were lower in the early season and increased with the advancing season. Duodenal flow of total, essential and non-essential amino acids also were increased with AO in late summer. In another metabolism study by the same group (Westvig, Caton and Erickson, 1993), steers were fed a basal forage diet with or without 0.9 kg barley 100 kg^{-1} bodyweight to which 0 or 2 g day^{-1} AO was added. The barley increased microbial protein synthesis and amino acid flow to the duodenum, but AO had no effect on performance or fermentation parameters.

In ruminally cannulated heifers, AO and YC^2 were tested at two fibre levels (27% and 37% NDF) in barley-corn starch diets (Miranda *et al.*, 1994). The fungal cultures resulted in increased *in situ* digestibility of NDF from alfalfa (at 48-h incubation) and higher ruminal propionate concentrations at both fibre levels. In a study with crossbred finishing steers fed high concentrate diets, Rush, Weichenthal and Van Pelt (1994) reported that neither yucca extract (YE) nor AO added to YE had a significant effect on performance or carcass characteristics, even though daily gains were increased 17% by the YE. However, experimental variability in this study was high.

In conclusion, performance results from addition of YC to diets for growing and fattening cattle were generally positive with increases in daily gains averaging 7.3%. Because feed intakes were not affected by yeast supplementation ($< 1\%$), efficiencies of feed utilization also were about 6% higher for cattle fed YC. The data show variable results from addition of AO cultures to high forage or pasture diets fed non-lactating cattle. Improvement in gains and feed efficiencies shown for cattle fed YC on high forage diets might relate to greater cellulolytic activity in their rumens. Cattle subjected to fescue endophyte toxicity did not benefit from YC supplementation, but those challenged with IBRV showed fewer days sick and higher feed intakes when YC was added to the diet. AO cultures did not aid in adaptation of cattle to high grain diets (to 28 days), but they resulted in improved performance after adaptation. The several studies to determine effects of the fungal cultures on *in situ* and *in vivo* digestibilities of nutrients in pasture diets have yielded variable results, depending on season of year and stage of maturity of pasture plants.

7.4 LACTATING COWS

7.4.1 Milk yield and dry matter intake (DMI) effects

A large number of studies have tested the addition of probiotics to diets for lactating cows. As with beef cattle, the principal materials tested have been cultures of *Saccharomyces cerevisiae* and *Aspergillus oryzae*. Additionally, effects of feeding a live bacterial culture also were reported. As shown in Table 7.2, results with fungal cultures in lactating cow diets have been quite variable, with some trials showing significant increases in milk production and other positive indications such as increased intakes. Also reported in some studies have been higher nutrient digestibilities, reduced rectal temperatures during heat stress, improved reproductive status, higher ruminal pH and a greater stability of ruminal fermentation. On the other hand, some studies have shown no effect or have been slightly negative. Generally, the fungal cultures have proven more effective in cows fed diets containing medium to large amounts of concentrate, or during early lactation. The reason for greater response under such conditions might be related to an increased need for maintaining rumen stability when feeding high levels of concentrate or because of alleviation of the stress that early lactation cows usually undergo. Moreover, widely differing farm environments with respect to nutrition, and microbial challenges might affect response of cows to fungal cultures and account for some of the observed variability.

This review concentrates on results reported from addition of fungal culture to lactation diets during the past 10 years. Several research groups have devoted considerable effort to elucidation of effects of these additives in dairy cattle, and results of these groups are discussed.

Workers at the Rowett Research Institute in Scotland (Williams *et al.*, 1991) fed 50% or 60% concentrate with different forages (straw, grass hay, alfalfa hay, alfalfa silage and corn silage) to early- and mid-lactation cows supplemented with or without YC^2. Mean milk yield for 10 comparisons slightly favoured the YC^2 groups (26.1 vs. 25.5 kg day^{-1}) and mean intake of DM was 3% higher (19.9 vs. 19.3 kg day^{-1}) for cows fed YC^2. However, cows in the two trials where 60% concentrate was fed exhibited a significant response to YC^2 (3.2 kg day^{-1}). Type of forage did not seem to influence the benefit derived from the additive.

Florida workers have reported five lactation studies where cows were fed fungal cultures (one with YC^1, two with YC^2 and two with AO). Compared with controls, milk yields averaged 4.5% higher for the two yeast treatments and 1% higher for the AO treatments. However, in the Florida AO trials, a number of additional feed ingredients were also

Table 7.2 Effect of fungal cultures (FC) on DMI and milk yield of lactating cows

Reference	No. cows	Diet	DMI (kg day^{-1}) −	+	Milk Yield (kg day^{-1}) −	+	Comment
Yeast culture studies							
Williams *et al.* (1991)	8	50% concentrate: 50% NH$_3$ trt. straw	15.7	16.5	21.7	20.6	Data for 7–12 weeks of lactation
	8	50% concentrate: ·50% grass hay	18.1	18.8	21.1	20.6	
	8	60% concentrate: 40% NH$_3$ trt. straw	17.8	18.7	19.4	22.8	
	8	60% concentrate: 40% grass hay	17.3	19.6	21.5	24.5	
Harris and Lobo (1988)	300	Corn silage and by product concentrates	NA	NA	26.6	27.6	Mid–late lactation
Harris and Smith (1992)	108	YC fed across alfalfa hay and corn silage diets	26.0	24.8	21.9	22.7	Experiment conducted for three 28-day periods
Harris and Webb (1990)	269	Corn silage: corn meal, and by products	26.8	25.9	30.9	31.7	Early–mid lactation
Arambel and Kent (1990)	20	31% alfalfa hay/ haylage: corn silage, barley, by products	29.1	29.0	36.3	35.5	Early lactation
Scott *et al.* (1994)	20	Not given	22.7	23.2	40.8	41.9	Early lactation
Kim *et al.* (1994)	30	Not given	NA	NA	30.0	28.2	Heifers in early lactation
Huber (1990)	520	Alfalfa silage: concentrate	NA	NA	28.4	29.2	Complete diet
Quinnez *et al.* (1988)		0.5 alfalfa hay: 0.5 corn	23.9	23.6	29.8	29.9	
		0.5 alfalfa hay: 0.5 wheat	21.8	21.5	28.4	29.1	
Bax (1988)		Grass silage: barley	—	—	18.8	21.1	
Dobos *et al.* (1990)		*Ad lib* silage:4 kg day^{-1} wheat: available pasture	—	—	26.4	28.6	

Table 7.2 Continued

Reference	No. cows	Diet	DMI (kg day^{-1})		Milk Yield (kg day^{-1})		Comment
			−	+	−	+	
Günther (1990)	35	0.24 maize silage: 0.34 grass silage:	—	—	34.4	40.4	Early/mid-lactation
		0.42 concentrate Silage, pasture, barley	—	—	31.9	36.6	Late lactation
Weiss (1991)	28				21.3	22.1	
Rodriguez-Salazar et al. (1991)	24	0.4 forage: 0.6 concentrate	Tended to eat more		20.6	22.5	
Wohlt et al. (1991)	24	0.5 corn silage–alfalfa hay: 0.5 grain	13.4	14.9	29.5	30.5	1–18 weeks lactation
Erdman and Sharma (1989)		0.4 corn silage: 0.6 concentrate	19.6	19.0	25.7	25.3	Mid-lactation
Putnam et al. (1994a)	4	0.44 forage: 0.56 concentrate	18.2	19.1	31.3	32.0	Primiparous cows
Swatz et al. (1994)	59	Field trial on 7 farms	Yeast had no effect		36.0	36.5	Early lactation
	247				31.8	31.8	All stages lactation
Shaver and Garrett (1995)	585	Field trial on 11 farms	—	—	36.7	37.6	150 day adjusted milk
Piva et al. (1993)	24	0.3 corn silage: 0.22 alfalfa hay: 0.48 concentrate	21.1	22.8	25.4	26.2	Mid-lactation

Aspergillus oryzae culture studies

Reference	No. cows	Diet	DMI (kg day^{-1})		Milk Yield (kg day^{-1})		Comment
			−	+	−	+	
Harris et al. (1983)	34	Silage diets	22.8	24.3	24.7	25.3	Mid-lactation
		Pelleted cotton-seed hulls multifactor diets	27.3	26.9	26.9	26.4	
Van Horn et al. (1984)	32	Non-forage fibre, multifactor diets	22.7	23.5	24.5	26.2	Early to mid-lactation
Wallentine et al. (1986)	50	High concentrate	—	—	30.1	33.6	Early lactation
	50	Maximum concentrate	—	—	29.0	27.8	Mid-lactation
Kellems et al. (1990)	100	High concentrate, complete ration, alfalfa silage	—	—	26.0	27.7	Complete lactation
Huber et al. (1985)	48	0.5 concentrate	18.6	19.4	18.5	20.4	Mid-lactation, primiparous cows
		0.3 concentrate	18.1	18.2	18.6	18.6	

Table 7.2 Continued

Reference	No. cows	Diet	DMI (kg day^{-1}) −	DMI (kg day^{-1}) +	Milk Yield (kg day^{-1}) −	Milk Yield (kg day^{-1}) +	Comment
Marcus *et al.* (1986)	205	High concentrate	—	—	28.7	30.2	Commerical herd (6 months)
Gomez-Alarcon *et al.* (1991)	24	0.5 concentrate	19.1	19.9	22.3	23.2	Mid-lactation
	36	0.6 concentrate	25.1	25.6	37.3	39.8	Early lactation
Denigan *et al.* (1992)	40	High concentrate,	22.6	23.7	30.9	30.8	Early lactation
	24	complete ration, alfalfa hay	25.5	24.7	26.2	25.8	Early lactation Mid-lactation
Yu *et al* (1994)	32	0.4 alfalfa: 0.6 concentrate	28.0	27.3	39.7	38.7	Early/mid-lactation
Sievert and Shaver (1993a)	6	Two NFC levels	23.3	23.5	30.9	31.1	Early lactation
Sievert and Shaver (1993b)	8	Two NFC levels	24.2	24.7	34.8	35.2	Early lactation
Higginbotham *et al.* (1993)	110	Field study	—	—	39.3	39.6	Commercial herd

DMI, dry matter intake
NFC, non-fibre carbohydrate; NA, not available.

being tested, which would interact with the effects of the fungal culture and make results difficult to interpret.

Three trials conducted by Utah State workers (one with YC[1] and AO, one with YC[2] and one with AO) showed small increases in milk yield from feeding YC[2] and slight decreases from feeding AO and YC[1]; whereas, workers at Brigham Young University (BYU), also located in Utah, reported an average of 5.8% increase in milk yields from feeding AO in two trials with five comparisons. Of particular note in both BYU studies was a considerably greater response in early- than mid-lactation cows.

Researchers at the University of Arizona (Huber *et al.*, 1994) conducted five lactation trials with AO and one with YC[2]. For the AO studies, three of eight comparisons showed significantly higher milk yields in cows fed the additive, and the other five were not significantly different from controls. The mean increase in milk yield was 2.5%. The cows fed the YC[2] (in a large field trial) also yielded more milk (28.4 vs. 29.2 kg day^{-1}) than controls ($P < 0.07$).

Greatest lactational response (38.5 vs. 33.2 kg day^{-1}) to a fungal additive (YC[2]) was reported by German workers (Günther, 1989, 1990), who fed early- and mid-lactation cows a diet of maize and grass silages plus 42% concentrate. Such large responses are atypical of reported data and difficult to explain. A number of other research groups have reported one or two trials which tested the fungal cultures with varied results; several have been positive, fewer were negative, and some were without effect. Effects on milk yields and intakes from studies where fungal cultures were fed to lactating cows are summarized in Table 7.2.

Wallace and Newbold (1992) explained the variation in lactation response to the fungal cultures on the basis of a possible improvement in a number of dietary properties which might cause benefit in specific situations, such as better fibre utilization on high concentrate diets because of greater numbers of cellulolytic bacteria, or increased rumen stability during early lactation. The increased rumen stability might explain the observation of Gomez-Alarcon *et al.* (1991), who reported a more rapid increase in feed intake immediately postpartum in cows fed 3 g day^{-1} of AO compared with controls.

Higher ruminal pH and enhanced metabolism of ruminal lactate, by both YC and AO, have been demonstrated by several groups (Nisbit and Martin 1989; Wallace and Newbold, 1992). Such effects might alleviate acidotic stress caused by high concentrate diets. Another factor that might have influenced the responses observed with fungal cultures is that practically all the studies were conducted in university herds where overall management and nutrition might be at a higher level than in many commercial herds. Hence, opportunity for cows to respond to some of the subtle beneficial effects associated with fungal cultures might be minimized.

Reduced acidotic stress might explain the results of Jahn, Chandler and Miller (1973) who fed a live inoculum of acetogenic bacteria (355 ml; 5×10^8 bacteria ml^{-1}) specifically adapted *in vitro* for efficient fermentation of starch to VFA. The inoculum treatment resulted in higher ruminal acetate and lower lactic acid concentrations. Compared with controls, inoculated cows produced 9% more milk (20.8 vs. 22.7 kg day^{-1}) with a consistently higher milk fat content, had similar feed intakes (17.9 vs. 17.7 kg day^{-1}) and gained less body weight (0.35 vs. 0.11 kg day^{-1}). A small elevation in molar percentage of acetate in the rumen was shown for inoculated cows.

Table 7.2 presents milk yield and DMI responses to fungal cultures, which have been divided into yeast and AO effects. Even though the studies did not always show increases in milk production and one cannot be guaranteed higher milk yields from use of fungal cultures, other benefits have been observed which might have been just as important to the dairy producer. Such benefits include that reported by

Robinson (1995), in which cows supplemented with yeast lost less body condition prepartum and exhibited numerically higher weight gains than unsupplemented cows. Moreover, the supplemented cows exhibited a higher calculated energy density vs. the NRC (1989) estimated value when compared with those not receiving yeast. Robinson's results did not confirm the finding of a more rapid rate of DMI increase postpartum with the yeast supplemented cows, previously reported by Gomez-Alarcon *et al.* (1991).

Milk yields for 31 comparisons (Table 7.2) of cows fed yeast culture vs. controls showed increased yields in 74% and a decrease in 22%, with an average advantage for treated cows of $1\,\mathrm{kg\,day^{-1}}$ or 3.6%. For the 17 comparisons of cows fed AO culture vs. controls, 65% were higher and 29% lower for the AO groups. The average yield increase for AO cows was $0.7\,\mathrm{kg\,day^{-1}}$.

Pertaining to feed intakes, mean values for 17 comparisons reporting DMI (Table 7.2) were 1% higher for cows fed YC compared with controls, and in AO-supplemented comparisons, dry matter intake (DMI) also averaged 1% more than the unsupplemented. These mean differences were negligible and do not suggest that the fungal cultures were improving the cows' willingness to consume their diets. When DMI differences between treated and controls were compared for those studies reporting at least $0.7\,\mathrm{kg}$ increase in milk yield from culture addition, the advantage in DMI for cows receiving fungal cultures was 2.5% or $0.5\,\mathrm{kg\,day^{-1}}$. The increased DMI in these comparisons would not usually supply sufficient energy to account for the higher milk production. This conclusion is in contrast with that of Williams and Newbold (1990), who proposed that no other physiological action need be proposed for the effects of fungal cultures other than the additional energy supplied by the extra dietary DM consumed. Certainly, greater cellulolysis, increased ruminal pH and accelerated lactate utilization, as well as reduced diurnal variation in ruminal metabolites (lactate, VFA, ammonia) might be associated with higher intakes. These effects have all been attributed to supplementation of the fungal cultures and could have benefited cows in ways other than a direct influence on intakes.

7.4.2 Milk component effects

Similar to milk yields, effects of fungal cultures on milk composition have been variable. Even though isolated studies have shown higher milk fat (Marcus, Huber and Cramer, 1986; Harris and Lobo, 1988; Günther, 1989) or milk protein (Günther, 1989; Higginbotham, Bath and Butler, 1993; Yu *et al.*, 1984), no trend towards consistent increases or decreases in percentages of milk fat, protein, lactose or solids non-fat (SNF) are supported by the several studies summarized. Huber *et al.*

(1994) reviewed effects of AO on milk composition which were reported in six studies. Results showed equal percentages of milk fat, milk protein and milk lactose. A similar review with the same results was made by Williams and Newbold (1990) for what was then the available data on effects of YC and AO on butterfat content of milk.

7.4.3 Heat stress effects

Heat-stressed cows fed AO in several studies had lower rectal temperatures and respiratory rates, or both, than companion controls (Gomez-Alarcon *et al.*, 1991; Higginbotham, Bath and Butler, 1993; Huber and Higginbotham, 1985; Marcus, Huber and Cramer, 1986), even though magnitude of effects have been variable (Table 7.3). In other trials, cows fed AO did not show decreased temperatures (Kellems, Lagerstedt and Wallentine, 1990; Denigan *et al.*, 1992), or they tended to be increased (Wallentine *et al.*, 1986) compared with controls. In one trial with a yeast culture (Huber, 1990), significant decreases in rectal temperatures were reported for cows fed the culture in the early stages of the study,

Table 7.3 Rectal temperatures (RT) and respiration rates (RR) of cows fed fungal cultures

Reference	Trial	RT		RR	
		Control	AO	Control	AO
Huber and Higginbotham	1	39.4	38.4[a]	—	—
(1985)	2	39.8	39.3[a]	—	—
Marcus *et al.* (1986)	1	40.1	39.9[c]	—	—
Gomez-Alarcon *et al.*	1	39.2	39.0[c]	67	63
(1991)	2	39.0	38.7[c]	—	—
	3[1]	38.7	38.5	—	—
Wallentine *et al.* (1986)	1	39.1	39.2[b]	79	82
	2	39.1	39.1	—	—
Kellems *et al.* (1990)	1	38.8	38.8	—	—
Denigan *et al.* (1992)	1	38.8	38.8	—	—
	2	39.3	39.3	70	74
Higginbotham *et al.* (1993)	1	38.8	38.7[a]	51	52

[a]Mean differences for all measurements lower for AO than control ($P < 0.05$).
[b]Mean differences of measurements in early lactation cows higher for AO than control ($P < 0.05$).
[c]About one-half of weekly means lower for AO than control ($P < 0.05$).
[1]Means of diurnal changes in inner ear for seven cows per treatment.
AO, *Aspergillus oryzae*.

but differences did not continue throughout the entire 6-month trial. No other studies were found pertaining to heat stress and yeast cultures in dairy cows. However, many of the other effects shown for AO cultures have also been demonstrated for yeast cultures (Huber, 1990; Williams and Newbold, 1990).

In an Arizona study, water intake was slightly higher for cows fed AO, and no effect on DMI was reported (Gomez-Alarcon *et al.*, 1991). It is not likely that the higher water intake would reduce body temperatures during heat stress, and the lack of difference in DMI suggests that heat of fermentation was not affected. A possible explanation for the reduced heat stress in cows fed fungal cultures is a specific influence of fungal metabolites on temperature control centres. Such an effect has been suggested by Meyers (1974), who reported that a number of compounds elaborated from fungi affect temperature control centres in animals. Typical of such an effect is the decreased body temperature observed in cows fed the aflatoxin from *Aspergillus flavus* (Mertens, 1979).

One might postulate that the enhanced nutrient utilization and increased milk yields that often occurred when fungal additives were fed would tend to increase, instead of decrease, body temperatures because of a tendency to cause higher intakes and greater milk yields. Increased rectal temperatures were observed in one study (Wallentine *et al.*, 1986) with cows that were not subjected to extremely high ambient temperatures over extended periods, as were cows in the Arizona studies. The cows in the BYU study which were fed AO early in lactation showed significantly higher milk yields than controls, but they also had higher rectal temperatures and respiration rates. On the other hand, milk yield, body temperatures and respiration rates of mid-lactation cows in the same study were not affected significantly by the additive. It is of interest, however, that the greatest increase in milk yields in the early lactation cows fed fungal culture occurred when environmental temperatures were highest (Wallentine *et al.*, 1986). Because cows were group fed, individual DMI could not be measured, but group intakes were the same for both treatments (21 kg day^{-1}).

As might be surmised, considerably more study is needed to clarify a relationship between feeding of fungal additives to dairy cows and their tolerance to heat stress. Further investigation should clearly establish whether such a relationship truly exists and should clarify the mechanism that might elicit such an effect.

7.5 MARKET PENETRATION OF PROBIOTICS

Very little information has been released on actual sales of probiotics or direct-fed microbials in cattle rations. The manufacturing companies are not prepared to release such figures, and usage surveys conducted with producers tend to be inaccurate because many producers are purchasing concentrate mixtures which may or may not contain one of several probiotics.

The *Hoard's Dairyman* magazine[*] has conducted an annual survey on use of various feed ingredients among their producer subscribers for several years. Approximately 1500 questionnaires were mailed, and about 600 were returned. For 1990, 23.2% of the respondents said they used yeast or yeast cultures, and 2.2% used what was termed 'probiotics'. However, there was no way of determining what 'probiotics' actually included. By 1993, usage of yeast and yeast cultures increased to 27.3% and probiotics to 8.6%. The 1993 figures are compared with 61.5% of dairy producers which stated they were feeding buffers, 27.9% niacin and 21.8% a zinc additive.

As stated previously, these usage estimates are somewhat lower than what is actually being fed because a number of feed manufacturers in the US and other countries are incorporating probiotic additives into prepared feed mixtures without the knowledge of the producer. Perhaps a more valid estimate of market penetration of yeast and yeast cultures is that of Hutjens (1993), who stated that 51% of the high-producing herds in the US were fed this additive in 1992 compared with 17% in 1983.

Hutjen's estimates are close to that of a manufacturer representative (not named) who suggested that 40% of US dairy cows are fed yeast. It was further stated by this representative that some of the transition diets for beef cattle entering feedlots carry yeast culture to help stabilize intakes early in the feeding period and that large amounts are being sold in a number of countries for beef cattle consuming mostly forage diets.

One observation from these reports which appears consistent is that usage of probiotics in dairy herds is steadily increasing. No information was available of probiotic usage on other types of cattle and calves, or on market penetration outside the US, but private communication and personal experience suggest that these sales also are rapidly increasing in other countries such as Mexico and Brazil.

7.6   HOW DO PROBIOTICS WORK IN CATTLE?

Reviews have been written on modes of action of direct-fed bacterial cultures in monogastrics which would be similar to liquid diets in baby

---

[*]Source was Steve Larson, Associate Senior Editor, *Hoard's Dairyman*, Ft Atkinson, Wisconsin.

calves, and on fungal and bacterial cultures in ruminants which take into account growing, feedlot and range cattle, as well as dairy heifers and cows (Wallace and Newbold, 1992). Effects of bacterial cultures in ruminants undergoing stress have also been reviewed (Males and Johnson, 1990). Therefore, this chapter only summarizes concepts already presented in other literature and gives judgement as to their importance and validity. Citing of the numerous separate reports written on mechanisms of probiotic action seems redundant.

7.6.1 Bacterial cultures

The excellent chapter by Wallace and Newbold (1992) in the book edited by Fuller cites several authors showing that lactobacilli and enterococci reduced diarrhoea in young ruminants, as has also been illustrated in this chapter. However, the exact mechanism(s) of the effects have not as yet been clarified. Possibilities proposed in several research reports are:

- adhesion to the gut wall to prevent colonization of coliforms;
- neutralization of enterotoxins (produced by an as yet unidentified metabolite);
- decrease in gut pH through production of organic acids suggested to inhibit excessive proliferation of coliforms (Fox, 1988);
- stimulation of synthesis of specific factors such as hydrogen peroxide capable of inhibiting the growth of potential pathogens.

One or a combination of these effects might also explain the favourable results obtained from feeding bacterial cultures to developing ruminants and cattle adapting to feedlot or other stressful situations.

7.6.2 Fungal cultures

Fungal culture effects on milk production and weight gains have been attributed primarily to increased DM intakes in numerous studies (Wallace and Newbold, 1992). However, the studies herein reviewed with growing cattle and milking cows show that mean intakes were not affected substantially, despite greater gains or milk yields. If intake is the major determinant of response in ruminants fed fungal cultures, then several factors observed to be affected by fungal cultures might influence intakes. These are rate and extent of digestion of dietary fibre, and rate of feed passage.

Several authors have shown that fungal cultures enhance a faster and more complete disappearance of fibre in *in vitro*, *in situ* and *in vivo* comparisons, but some have failed to repeat this effect (Wallace and Newbold, 1992). Explanations for the increased fibre digestibility might be increased numbers of cellulolytic bacteria, improved ruminal metabolism resulting from increased concentrations of lactic acid fermenting bacteria (*Selenomonas ruminantium*), and higher ruminal pH. These effects have all been attributed to addition of fungal cultures into rumen fermentation systems. The chapter by Wallace and Newbold (1992) tends to favour stimulation of cellulolytic bacteria as the most logical explanation for increased cellulose disappearance, which would ultimately impact intakes. They discount the possibility that higher pH or decreased lactate concentrations are responsible for the increased fibre-digesting bacteria because the magnitude of these changes is small, even though they admit that they may have some influence on growth of cellulolytic bacteria. These authors suggest that possible reasons for the increases in cellulolytic bacteria are removal from the media of sugars, toxic metal ions or molecular oxygen – all of which might inhibit their growth. Stimulation of bacterial growth might also be affected by the presence in yeast and AO of cellulase enzymes or specific nutrients or cofactors which may enhance bacterial growth. The scheme (Figure 7.1) proposed by Wallace and Newbold for explaining

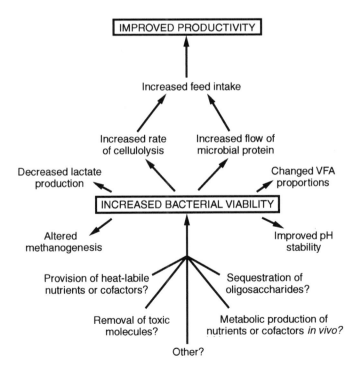

Fig. 7.1 An alternative scheme for investigating the primary mode of action of fungal feed additives. (Proposed by Wallace and Newbold, 1992).

fungal culture effects in ruminants shows primary and secondary events affecting animal performance. As noted, many unanswered questions remain and the validity of a number of these effects is, as yet, unproven.

The question of whether the yeast and AO populations remain viable and multiply in the rumen has often arisen. Most studies suggest that because of the anaerobic environment of the rumen, there is no multiplication of either of these aerobic organisms after supplementation. Furthermore, viable counts usually decline to those found in control animals within 8–12 h after feeding. Hence, continued intake of fungal cultures is necessary to elicit an effect. Maintaining viable organisms of these fungal cultures for supplementation is important because it has been shown that autoclaving of both destroyed their effectiveness in stimulating bacterial growth in rumen fermentations, even though some activity of AO was maintained after gamma-radiation (Wallace and Newbold, 1992). An interesting and perhaps important finding is that suggested by Huber (1990) and confirmed by Welch and Calza (1993), who showed that the AO culture stimulated growth of normal rumen fungi (*Neosallimstis frontalis*) by about 20%. They interpreted the data to mean that part of the AO effect in ruminants might be mediated through changes in fungal physiology of the rumen.

7.7 CONCLUSIONS

The considerable data presented herein from numerous experiments have demonstrated an overall benefit from addition of probiotics to cattle diets. Despite the considerable variation in the studies, careful weighing of the advantages and disadvantages of using probiotics for the various classes of livestock would lead one to conclude that the pros outweigh the cons.

To the question often asked by producers and scientists 'Do they work?', it is not possible to provide a simple answer. Often more than just production responses must be taken into consideration to portray the real value of probiotics in cattle rations. For example, feeding of bacterial-type inocula to baby calves may elicit more rapid starter intake or decrease incidence of diarrhoea during a particular part of the preweaning period, but may not affect overall weight gains or feed intake. Such an effect may show slight differences in mortality rates, but trials often do not contain a sufficient number of animals to show significant effects. Because they may act on a multiplicity of organisms at different sites of the digestive tract, the different probiotics may have multiple actions and some may be additive or negative in relation to others.

Greatest usage of probiotics, particularly the fungal cultures, has been with high-forage diets in beef cattle and with early lactation, high concentrate diets in dairy cattle. These usage patterns reflect a three-fold higher response of beef cattle to YC when high forage vs. high concentrate diets were fed. In beef studies, greater nutrient utilization has been attributed to probiotic addition, probably because of stimulation of higher numbers of fibre-digesting bacteria.

In dairy cattle, mean increases in milk yields of 0.7–1.0 kg day^{-1} have resulted from addition of fungal cultures to diets, with two-thirds of 48 studies reporting positive results. As mentioned, responses were greater during early lactation when cows were undergoing postpartum stresses. Reasons for benefits from feeding the probiotics mentioned by various authors have been more stable rumen conditions, leading to more rapid increases in intakes and less digestive disturbances; more rapid turnover of ruminal lactic acid and a higher ruminal pH; lower rectal temperatures during heat stress conditions (though not confirmed in some studies); and higher overall intakes of feed (though mean intakes differed little).

A steady increase in purchase of probiotics has been shown over the last decade as illustrated by estimates of as many as 50% of the dairy producers in the US using YC. With an estimated response of 1 kg day^{-1} in milk yield and very little increase in feed consumption, the cost associated with using the material would be only its purchase price. Hence, a benefit to cost ratio for yeast culture under US conditions would be 4 or 5:1, which, under most conditions, is very favourable.

7.8   REFERENCES

Adams, D.C., Galyean, M.L., Kiesling, H.E. *et al.* (1981) Influence of viable yeast culture, sodium bicarbonate and monensin on liquid dilution rate, rumen fermentation and feedlot performance of growing lambs and digestibility in lambs. *J. Anim. Sci.*, **53**, 780–9.

Allison, B.C. and McCraw, R.L. (1989) Efficacy of Vita-ferm formula for stocker calves. *Anim. Sci. Newsletter A&T and NC State Uni.*, November, pp. 12–18.

Anderson, V.L. (1993) Diets with and without yeast for early weaned beef calves. *J. Anim. Sci.*, **71** (Suppl. 1), 239.

Arambel, M.J. and Kurt, B.A. (1990) Effect of yeast culture on nutrient digestibility and milk yield response in early- to midlactation dairy cows. *J. Dairy Sci.*, **73**, 1560–3.

Bax, J.A. (1988) An investigation into the response of dairy cows to supplementation with yeast culture, in *2nd Intl Symp. New Techniques in Ag (New Techniques in Cattle Production)* (ed. C.J.C. Phillips), University College of North Wales, U.K.

Bechman, T.J., Chambers, J.V. and Cunningham, M.D. (1977) Influence of *Lactobacillus acidophilus* on performance of young dairy calves. *J. Dairy Sci.*, **60** (Suppl. 1), 74.

Beckett, J.L. and Oltjen, J.W. (1992) Estimation of the water requirement for beef production in the United States. *J. Anim. Sci.*, **70** (Suppl. 1), 287.

Beeman, K. (1985) The effect of *Lactobacillus* spp. on convalescing calves. *Agripractice*, **6**, 8–10.

Beharka, A.A., Nagaraja, T.G. and Morrill, J.L. (1990) Ruminal microbial and metabolic development in young calves fed calf starter supplemented with *Aspergillus oryzae* extract. *J. Dairy Sci.*, **73** (Suppl. 1), 220.

Bonaldi, G., Buratto, L., Darsie, G. *et al.* (1986) Use of *Saccharomyces cerevisiae* and *Lactobacillus acidophilus* in veal calves. *Obiet. Document. Vet.*, **7**, 49–53.

Bruce, B.B., Gilliland, S.E., Bush, L.J. and Staley, T.E. (1979) Influence of feeding cattle cells of *Lactobacillus acidophilus* on faecal flora of young dairy calves. *Oklahoma Anim. Sci. Res. Rep.*, **207**.

Caton, J.S., Erickson, D.O., Carey, D.A. and Ulmer, D.L. (1993) Influence of *Aspergillus oryzae* fermentation extract on forage intake, site of digestion, *in situ* degradability, and duodenal amino acid flow in steers grazing cool-season pasture. *J. Anim. Sci.*, **71**, 779–87.

Coffey, K.P., Brazle, F.K., Moyer, J.L. *et al.* (1990) Performance of stocker steers offered fescue or bromegrass screenings with and without *Aspergillus oryzae* fermentation extract. *J. Anim. Sci.*, **68** (Suppl. 1), 574.

Coffey, K.P., Brazle, F.K. Ruehlow, R.A. and Moyer, J.L. (1993) Digestibility of *Acremonium coenophialum*-infected and endophyte-free fescue hay diets supplemented with *Aspergillus oryzae* fermentation extract. *J. Anim. Sci.*, **71** (Suppl. 1), 30.

Cole, N.A., Purdy, C.W. and Hutcheson, D.P. (1992) Influence of yeast culture on feeder calves and lambs. *J. Anim. Sci.*, **70**, 1682–90.

Denigan, M.E., Huber, J.T., Alhadhrami, G. and Al-Dehneh, A. (1992) Influence of feeding varying levels of Amaferm® on performance of lactating dairy cows. *J. Dairy Sci.*, **75**, 1616–21.

Dobos, R.C., Dickens, A.J. and Norris, T.L. (1990) Yea-Sacc[1026] for dairy cattle in low concentrate input systems: effects on milk yield and composition in an Australian experiment, in *Biotechnology in the Feed Industry* (ed. T.P. Lyons), Alltech Technical Publications, Nicholasville, Kentucky, pp. 518–19.

Eckles, C.H., V.M. Williams and J.W. Wilber (1924) Yeast as a supplementary feed for calves. *J. Dairy Sci.*, **7**, 421.

Edwards, I.E., Mutsvangwa, T., Topps, J.H. and Paterson, G.F.M. (1990) The effect of supplemental yeast culture (Yeasacc) on patterns of rumen fermentation and growth performance of intensively fed bulls. *Anim. Prod.*, **51**, 579.

Ellinger, D.K., Muller, L.D. and Glantz, P.J. (1978) Influence of feeding fermented colostrum and *Lactobacillus acidophilus* on faecal flora and selected blood parameters of young dairy calves. *J. Dairy Sci.*, **61** (Suppl. 1), 126.

Ely, D.G., Ragland, K.K., Johns, J.T. and Aaron, D.K. (1990) Amaferm supplementation to steers grazing tall fescue. *J. Anim. Sci.*, **68** (Suppl. 1), 31.

Fallon, R.J. and Harte, F.J. (1987) The effect of yeast culture inclusion in the concentrate diet on calf performance. *J. Dairy Sci.*, **70** (Suppl. 1), 119.

Fox, S.M. (1988) Probiotics: intestinal inoculants for production animals. *Vet. Med.*, **August**, 806–30.

Gilliland, S.E., Bruce, B.B., Bush, L.J. and Staley, T.E. (1980) Comparison of two strains of *Lactobacillus acidophilus* as dietary adjuncts for young calves. *J. Dairy Sci.*, **63**, 964–72.

Gomez-Alarcon, R.A., Huber, J.T., Higginbotham, G.E. *et al.* (1991) Influence of feeding an *Aspergillus oryzae* culture on the milk yields, eating patterns, and body temperatures of lactating cows. *J. Anim. Sci.*, **69**, 1733–40.

Günther, K.D. (1989) Yeast culture's success under German dairy conditions, in

Biotechnology in the Feed Industry (ed. T.P. Lyons), Alltech Technical Publications, Nicholasville, Kentucky, pp. 39–46.

Günther, K.D. (1990) Yeast culture: success under German conditions. *Feed Compounder*, January.

Harris, B. and Lobo, R. (1988) Feeding yeast culture to lactating dairy cows. *J. Dairy Sci.*, **71** (Suppl. 1), 276.

Harris, B. and Smith, W.A. (1992) Feeding dairy cattle in early lactation for maximum efficiency, in *Biotechnology in the Feed Industry* (ed. T.P. Lyons), Alltech Technical Publications, Nicholasville, Kentucky, pp. 99–107.

Harris, B., Jr., Van Horn, H.H., Manookian, K.E., *et al.* (1983) Sugarcane silage, sodium hydroxide and steam pressure-treated sugarcane bagasse, corn silage, cottonseed hulls, sodium bicarbonate, and *Aspergillus oryzae* product in complete rations for lactating cows. *J. Dairy Sci.*, **66**, 1474–85.

Harris, B. and Webb, D.W. (1990) The effect of feeding a concentrated yeast culture product to dairy cows. *J. Dairy Sci.*, **73** (Suppl. 1), 266.

Hefel, H. (1980) Cited by Fox, S.M. (1988) Probiotics: intestinal inoculants for production animals. *Vet. Med.*, August, 806–30.

Higginbotham, G.E. and Bath, D.L. (1993) Evaluation of *Lactobacillus* fermentation cultures in calf feeding systems. *J. Dairy Sci.*, **76**, 615–20.

Higginbotham, G.E., Bath, D.L. and Butler, L.J. (1993) Effect of feeding *Aspergillus oryzae* extract on milk production and related responses in a commercial dairy herd. *J. Dairy Sci.*, **76**, 1484–9.

Huber, J.T. (1990) The fungal and yeast culture story in lactating dairy cows, in *Proc. Southwest Nutr. Manage. Conf.*, Tempe, AZ, pp. 87–94.

Huber, J.T. and Higginbotham, G. (1985) Influence of feeding vitaferm, containing an enzyme-producing culture from *Aspergillus oryzae*, on performance of lactating cows. *J. Dairy Sci.*, **68** (Suppl. 1), 122.

Huber, J.T., Higginbotham, G., Gomez-Alarcon, A. *et al.* (1994) Heat stress interactions with protein, supplemental fat, and fungal cultures. *J. Dairy Sci.*, **77**, 2080–90.

Hudyma, W.T. and Gray, D. (1993) Effect of feeding yeast culture and sorting calves by weight on feed-lot performance of calves fed a corn silage diet. *J. Anim. Sci.*, **68** (Suppl. 1), 471.

Hughes, J. (1988) The effect of a high strength yeast culture in the diet of early weaned calves. *Anim. Prod.*, **46**, 526.

Hutjens, M.F. (1993) Feed additives. *Proc. Western Large Herd Manage. Conf.*, Las Vegas, NV, pp. 168–73.

Jahn, E., Chandler, P.T. and Miller, C.N. (1973) Lactational response of dairy cows inoculated with live adapted rumen microorganisms. *J. Anim. Sci.*, **56** (Suppl. 1), 643.

Kellems, R.O., Lagerstedt, A. and Wallentine, M.V. (1990) Effect of feeding *Aspergillus oryzae* fermentation extract or *Aspergillus oryzae* plus yeast culture plus mineral and vitamin supplement on performance of Holstein cows during a complete lactation. *J. Dairy Sci.*, **73**, 2922–8.

Kim, D.Y., Dawson, D.P., Kent, B.A. and Arambel, M.J. (1994) Effect of supplemental viable yeast culture with or without *Aspergillus oryzae* on body weight gain, milk production and nutrient digestibility in early lactating Holstein heifers. *J. Anim. Sci.*, **72** (Suppl. 1), 299.

Lee, R.W. and Botts, R.L. (1988) Evaluation of single oral dosing and continuous feeding of *Streptococcus faecium* M74 (syntabac) on performance of incoming feedlot cattle. *J. Anim. Sci.*, **66** (Suppl. 1), 460.

Maeng, W.J., Kim, C.W. and Shin, H.T. (1987) Effect of a lactic acid bacteria concentrate (*Streptococcus faecium* Cernelle 68) on growth rate and scouring prevention in dairy calves. *Korean J. Dairy Sci.*, **9**, 204–10.

Males, J.R. and Johnson, B.J. (1990) Probiotics: What are they? What do they do? *J. Anim. Sci.*, **68** (Suppl. 1), 504.

Marcus, K.M., Huber, J.T. and Cramer, S. (1986) Influence of feeding Vitaferm during hot weather on performance of lactating cows in a large dairy herd. *J. Dairy Sci.*, **69** (Suppl. 1), 188.

Mertens, D.R. (1979) Biological effects of mycotoxins upon rumen function and lactating dairy cows, in *Interaction of Mycotoxins in Animal Production*, Natl Acad. Sci., Washington, DC, pp. 118–36.

Meyers, R.D. (1974) *Handbook of Drug and Chemical Stimulation of the Brain*, Van Nostrand Reinhold, New York, NY.

Mir, Z. and Mir., P.S. (1994) Effect of the addition of live yeast (*Saccharomyces cerevisiae*) on growth and carcass quality of steers fed high-forage or high-grain diets and on feed digestibility and *in situ* degradability. *J. Anim. Sci.*, **72**, 537–45.

Miranda, R.L.A., Mendoza, G., González, M. and Sárcena, R. (1994) Effect of *Saccharomyces cerevisiae* and *Aspergillus oryzae* cultures and NDF level on ruminal fermentation. *J. Anim. Sci.*, **72** (Suppl. 1), 1152.

Morrill, J.L., Morrill, J.M., Feyerherm, A.M. and Laster, J.F. (1995) Plasma proteins and a probiotic as ingredients in milk replacer. *J. Dairy Sci.*, **78**, 902–7.

National Research Council (NRC) (1989) *Nutrient Requirements of Dairy Cattle*, 6th rev. ed., Natl Acad. Sci, Washington, DC.

Nisbet, D.J. and Martin, S.A. (1989) Factors affecting lactate uptake by *Slenomous rumateum*. Abstr. 20th Conf. on Rumen Fermentation, Chicago, IL.

Olson, K.C., Caton, J.S., Kirby, D.R. and Norton, P.L. (1994a) Influence of yeast culture supplementation and advancing season on steers grazing mixed-grass prairie in the Northern Great Plains. I. Dietary composition, intake, and *in situ* nutrient disappearance. *J. Anim. Sci.*, **72**, 2149–57.

Olson, K.C., Caton, J.S., Kirby, D.R. and Norton, P.L. (1994b) Influence of yeast culture supplementation and advancing season on steers grazing mixed-grass prairie in the Northern Great Plains. II. Ruminal fermentation, site of digestion, and microbial efficiency. *J. Anim. Sci.*, **72**, 2138–70.

Ozawa, K., Yabu-uchi, K., Yamanaka, K. *et al.* (1983) Effect of *Streptococcus faecium* BIO-4R on intestinal flora of weanling piglets and calves. *Appl. Environ, Microbiol.*, **45**, 1513–18.

Parra, I. and DeCostanzo, A. (1992) Influence of yeast culture supplementation during the initial 23 d on feed, or throughout a 58-d feeding period on performance of yearling bulls. *J. Anim. Sci.*, **70** (Suppl. 1), 287.

Piva, G., Belladona, S., Fusconi, G. and Sicbaldi, F. (1993) Effects of yeast on dairy cow performance, ruminal fermentation, blood components, and milk manufacturing properties. *J. Anim. Sci.*, **76**, 2717–22.

Putnam, D.E., Schwab, C.G., Socha, M.T. *et al.* (1994a) Effect of yeast culture on ruminal fermentation, nutrient digestibilities and passage, blood metabolites, and milk yield. *J. Anim. Sci.*, **72** (Suppl. 1), 188.

Putnam, D.E., Schwab, C.G., Socha, M.T. *et al.* (1994b) Intake and flow of N fractions and amino acids to the small intestine in primiparous Holstein cows fed supplemental yeast culture. *J. Anim. Sci.*, **72** (Suppl. 1), 242.

Quinonez, J.A., Bush, L.A., Nalsen, T. and Adams, G.D. (1988) Effect of yeast culture on intake and production of dairy cows fed high wheat rations. *J. Dairy Sci.*, **71** (Suppl. 1), 275.

Robinson, P.H. (1995) Effect of yeast on the adaptation of cows to diets postpartum. *J. Dairy Sci.*, **78** (Suppl. 1), 267.

Rodriguez-Salazar, O., Herrera-Saldana, R., Gonzalez-Munoz, S. and Miranda-Romero, L.A. (1991) The effect of Yea-Sacc[1026] (*Saccharomyces cerevisiae*) on dry matter degradability and milk production. *J. Anim. Sci.*, **74** (Suppl. 1), 105.

Rush, I.G., Weichenthal, B.A. and Van Pelt, B.G. (1994) Yucca extract with and without *Aspergillus oryzae* for finishing steers. *J. Anim. Sci.*, **72** (Suppl. 1), 1152.

Scott, S.K., Arambel, M.J., Kim, D.Y. *et al.* (1994) Effect of feeding yeast culture on milk production, composition, feed intake, and nutrient digestibility in lactating cows. *J. Anim. Sci.*, **72** (Suppl. 1), 300.

Seymour, W.M., Nocek, J.E. and Siciliano-Jones, J. (1995) Effects of a colostrum substitute and of dietary brewer's yeast on health and performance of dairy calves. *J. Dairy Sci.*, **78**, 412–20.

Shaver, R.D. and Garrett, J.E. (1995) Lactation responses to dietary yeast culture on commercial dairies. *J. Anim. Sci.*, **73** (1), 54.

Sievert, S.J. and Shaver, R.D. (1993a) Carbohydrate and *Aspergillus oryzae* effects on intake, digestion, and milk production by dairy cows. *J. Dairy Sci.*, **76**, 245–54.

Sievert, S.J. and Shaver, R.D. (1993b) Effect of nonfiber carbohydrate level and *Aspergillus oryzae* fermentation extract on intake, digestion, and milk production in lactating dairy cows. *J. Anim. Sci.*, **71**, 1032–40.

Svozil, B., Danek, P., Kumprecht, I. and Zobac, P. (1987) The efficiency of different contents of the bacterium *Streptococcus faecium* M-74 in the nutrition of calves. *Zivocisna Vyroba*, **32**, 265–71.

Swartzm, D.L., Muller, L.D., Rogers, G.W. and Varga, G.A. (1994) Effect of yeast cultures on performance of lactating dairy cows: a field study. *J. Dairy Sci.*, **77**, 3073–80.

Tournut, J. (1989) Applications of probiotics to animal husbandry. *Rev. Sci. Tech. Off. Int. Epiz.*, **8**, 551–6.

Van Horn, H.H., Harris, B., Taylor, M.J. *et al.* (1984) By-product feeds for lactating dairy cows: effects of cottonseed hulls, sunflower hulls, corrugated paper, peanut hulls, sugarcane bagasse and whole cottonseed with additives of fat, sodium-bicarbonate and *Aspergillus oryzae* product on milk production. *J. Dairy Sci.*, **67**, 2922–38.

Wallace, R.J. and Newbold, C.J. (1992) Probiotics for ruminants, in *Probiotics: The Scientific Basis* (ed. R. Fuller), Chapman & Hall, London, pp. 317–53.

Wallentine, M.V., Johnson, N.P., Andrus, D. *et al.* G.E. (1986) The effect of feeding *Aspergillus oryzae* culture-vitamin mix on the performance of lactating dairy cows during periods of heat stress. *J. Dairy Sci.*, **69** (Suppl. 1), 189.

Weiss, D. (1991) Effect of Yea-Sacc[1026] on milk yield and composition at a German dairy, in *Biotechnology in the Feed Industry* (ed. T.P. Lyons), Alltech Technical Publications, Nicholasville, Kentucky, p. 305.

Westvig, R.M., Caton, J.S. and Erickson, D.O. (1993) Influence of barley and *Aspergillus oryzae* fermentation extract supplementation on microbial efficiency, duodenal crude protein and amino acid flows, and digesta kinetics in steers fed prairie hay. *J. Anim. Sci.*, **71** (Suppl. 1), 81.

Wiedmeier, R.D. (1989) Optimizing the utilization of low quality forages through supplementation and chemical treatment. *Utah Beef Cattle Field Day*, **9**, 17–21.

Wiedmeier, R.D., Arambel, M.J. and Walters, J.L. (1987) Effect of yeast culture and *Aspergillus oryzae* fermentation extract on ruminal characteristics and nutrient digestibility. *J. Dairy Sci.*, **70**, 2063–8.

Williams, P.E.V. and Newbold, C.J. (1990) Rumen probiosis: the effects of novel microorganisms on rumen fermentation and ruminant productivity, in *Recent Advances in Animal Nutrition 1990* (eds W. Haresign and D.J.A. Cole), Butterworth, London, pp. 211–27.

Williams, P.E.V., Tait, C.A.G., Innes, G.M. and Newbold, C.J. (1991) Effects of the inclusion of yeast culture (*Saccharomyces cerevisiae* plus growth medium) in the diet of dairy cows on milk yield and forage degradation and fermentation patterns in the rumen of sheep and steers. *J. Anim. Sci.*, **69**, 3016–26.

Wohlt, J.E., Finkelstein, A.D. and Chung, C.H. (1991) Yeast culture to improve intake, nutrient digestibility and performance in dairy cattle during early lactation. *J. Dairy Sci.*, **74**, 1395–400.

Wren, B. (1987) Probiotics – fact or fiction. *Large Anim. Vet.*, Nov/Dec., 28–30.

Yu, P., Huber, J.T., Theurer, C.B. *et al.* (1994) Effect of feeding Amaferm® during hot weather on performance of dairy cows fed diets of steam-flaked or steam-rolled corn. *J. Anim. Sci.*, **77** (Suppl. 1), 383.

CHAPTER 8

Intervention strategies: the use of probiotics and competitive exclusion microfloras against contamination with pathogens in pigs and poultry

R.W.A.W. Mulder, R. Havenaar and J.H.J. Huis in 't Veld

8.1 INTRODUCTION

Since Metchnikoff's publication, *The Prolongation of Life* (Metchnikoff, 1908), at the beginning of this century there has been general agreement that the intestinal microflora has a strong effect on the health status of men and animals. In the 1970s and 1980s a great deal of effort was expended to improve the growth and health status of production animals by modulating the indigenous intestinal flora by live microbial adjuncts – probiotics. In these years several interesting publications appeared but consistent positive results in well-controlled double-blind studies were scarce. As a consequence, in the 1990s the interest in probiotics in animal nutrition has declined, especially in Europe. In contrast, research in human probiotics has received much attention and evidence is accumulating from well-designed, randomized

Probiotics 2: Applications and practical aspects.
Edited by R. Fuller.
Published in 1997 by Chapman & Hall.   ISBN 0 412 73610 1

and placebo-controlled double-blind studies that a few well-characterized lactic acid bacteria possess documented health-promoting effects (Lee and Salminen, 1995; Salminen *et al.*, 1996). The mechanisms behind the specific benefits include aspects such as immune modulation and strengthening the gut mucosal barrier due to:

- gut microflora modification;
- adherence to intestinal mucosa with capacity to prevent pathogen adherence;
- modification of dietary proteins by intestinal microflora;
- modification of bacterial enzyme capacity especially in relation to tumour formation;
- influence on gut mucosal permeability.

As a consequence, several probiotic supplements for humans have been introduced into the market.

Most of the above-mentioned mechanisms have been elucidated in diseased persons or persons with intestinal disorders. One can be sure that the expected effects are most prominent in such populations. So far no studies have been published in healthy volunteers where effects would be more difficult to demonstrate. Probiotics in human nutrition must, therefore, be considered as a way of maintaining a good health status and, possibly, in the long run, a way of preventing (chronic) diseases.

Several of the above-mentioned aspects are also of importance in improving the health status of production animals. Modern animal production practices do not promote a balanced intestinal microflora and metabolism. However, the interest of the feed industry in probiotics has decreased. Economic pressure on the meat industry and thus the lack of funds for fundamental and applied research and the observed variability in the results are probably the main reasons for this. But also in normal, healthy animals, an optimal balance of micro-organisms in the intestinal tract may contribute to resistance to infectious diseases and help in efficient digestion, especially under stress situations such as in modern meat production, which generally do not support the development and maintenance of a well-balanced intestinal microflora. As a consequence colonization with potential pathogenic micro-organisms such as *Salmonella, Campylobacter* or virulent *Escherichia coli* serotypes (e.g. K88, K99, 0157, H7) can occur. This puts even more pressure on the meat industry. Subsequently, the increased use of antibiotics in animal production has caused severe problems in both veterinary and human medicine. Can probiotics break this vicious circle?

Several probiotics for use in animals have been introduced into the market, but many of them soon disappeared again. All of them claimed to improve the performance and health status of production animals through an effect on the gut epithelium or on the gut flora.

How can probiotics improve digestibility and prevent colonization of zoonotic micro-organisms? The purpose of this chapter is to examine which of the various groups of probiotics are potentially beneficial for pigs and poultry by examining their possible mode(s) of action and to review some of the recent results obtained in practice.

8.2 INTESTINAL FLORA AND GUT METABOLISM

Sissons (1989) identified several mechanisms by which fat absorption and protein and carbohydrate digestion might be influenced by a change in the microflora of the gut. However, is it likely that a small shift in the microbial population due to application of probiotics could have a sustained effect on either uptake or nutrient degradability? As yet, the link between microbial load and the metabolic status of the gut has been little studied although this is changing (Wick *et al.*, 1991). Indeed, the gut represents only a small proportion of the total body mass but it must be realized that it is the single most demanding organ in the body in terms of its need for energy and protein. Thus any change in the metabolic demands of the gut is likely to have repercussions for the whole animal. It is well recognized that the indigenous microflora has a pronounced effect on the structure, function and (protein) metabolism of the gut (Miller *et al.*, 1986; White and Savage, 1987; Nouusiainen and Suoni, 1991). The physiological effects in relation to changes in microbial load and composition in pigs have been summarized as follows (Chesson, 1994):

- a reduction in protein turnover and energy requirements in the gut as a consequence of lower rates of crypt cell proliferation and reduced gut mass;
- a small reduction in protein demand in the liver due to changed immunological status;
- an increase in the proportion of amino acid pool available to other tissues, particularly for the synthesis of skeletal muscle;
- a reduction in the amount of nitrogen lost to the animal through leakage, mucin secretion and the imperfect recovery of protein sloughed off into the gut lumen;
- an increase in apparent nitrogen digestibility and absorption.

Furthermore the application of probiotics has been reported to increase levels of sucrase, lactase and tripeptidase activities associated with the brush border of the intestinal villi (Collington *et al.*, 1990).

Opportunistic pathogens exist as part of the microflora of the gut of pigs and poultry with no evident consequences for the host. Chronic and acute symptoms only become apparent when an overgrowth of these micro-organisms in, or spreading to, the small intestine is triggered. When this occurs, erosion of the villi and enterotoxin-induced systemic effects may develop. A reasonable number of harmless, autochthonous micro-organisms in close proximity to, or adherent on, the mucosal surface of the small intestine might trigger a cell-mediated immune response and could induce a more rapid rate of crypt cell proliferation. An increased rate of enterocyte migration up to the villi results in a defence reaction which allows the shedding and replacement of cells to which harmful micro-organisms are attached (Chesson, 1994).

If most of the nutrient sparing is a consequence of reducing the numbers or changing the nature of the adherent bacterial population then benefits seen in terms of animal performance are most likely in those animals with the poorest health status or at farms with poor hygiene management. This may be difficult to detect in animals living in herds. Even when effective with animals most at risk, the effects of probiotics may be diluted by the more limited response of the 'healthier' animals and, in herd terms, seem to be largely ineffective. It is, therefore, difficult to detect beneficial effects in herds of pigs and poultry living under normal conditions of animal production. This aspect has certainly influenced the acceptability of probiotics in animal feeding in the past.

8.3 MICROFLORA AND COLONIZATION RESISTANCE

The gastrointestinal tract of animals harbours a microflora which is formed immediately after the animal is born. The mature indigenous microflora forms an important barrier against colonization of potentially pathogenic micro-organisms such as *Salmonella* and *E. coli*. The microflora of the healthy animal consists of many different species of micro-organisms. *Lactobacillus*, *Bifidobacterium* and *Bacteroides* species are the most predominant groups of micro-organisms present. They constitute about 90% of the flora. Some other groups present include Enterobacteriaceae, *Enterococcus* and *Clostridium*. In healthy animals the ratio of numbers of micro-organisms of the different groups is important.

It is known that the microflora in the gastrointestinal tract of animals can be influenced, for example by ingestion of other micro-organisms. The results of such a treatment can be a change in number of micro-organisms already present, producing a less favourable environment for colonization, at a later stage, by potentially pathogenic micro-organisms. In the case of pigs, the administration of one or two different species of micro-organisms can exert a beneficial effect, but with artificially hatched poultry a microflora consisting of more than 30 different micro-organisms seems to be necessary.

Among the potential pathogens *Salmonella* and *Campylobacter* are considered to be the most serious causes of problems in poultry and *E.*

coli in pigs (host pathogenic strains as well as human pathogenic strains). As a result of modern husbandry practices, the treatment with vaccines and antimicrobials may result in a more favourable environment for colonization of these pathogens. The mechanisms of colonization by micro-organisms, as well as the occurring metabolic interactions in the gastrointestinal tract, are not well understood and future research should concentrate on these aspects, as intervention strategies will never be successful when basic knowledge is not available.

Factors known to influence the colonization of micro-organisms, although their quantitative contribution is not clear, can be grouped as follows:

- host-related factors (e.g. body temperature, pH and redox-potential levels, bile acids and enzymes, genetic resistance in different breeds);
- microbe-related factors (e.g. effects of antagonistic micro-organisms, bacteriophages, bacteriocins);
- diet- and environment-related factors (e.g. use of mannose, lactose and other (mixtures of) carbohydrates and/or dietary fibres as well as stress factors resulting from conditions at the farm level).

Therefore, the use of probiotic and competitive exclusion microflora products should exert an effect on factors as mentioned above. However, the knowledge of mechanisms is missing and therefore it follows that results of these treatments cannot be predicted and it can be expected that treatments will not be very effective, especially not under practical farm conditions.

The success of competitive exclusion microfloras against *Salmonella* and *Campylobacter* contamination in poultry in particular, even under practical farm conditions, has been described in the literature. However, a zero level of contamination could not be reached. From this work it became clear that *Lactobacillus* spp. and some other (mainly anaerobic) micro-organisms are very important in establishing a stable gastrointestinal microflora which is able to prevent colonization of potentially pathogenic micro-organisms. Therefore, the use of probiotic products containing *Lactobacillus, Bacillus* spores and yeasts may also exert a positive effect against this colonization.

Examples of results of trials with probiotic and competitive microflora products given to pigs and poultry under practical conditions are presented.

8.4 USE OF PROBIOTICS

Probiotics are defined as 'mono- or mixed cultures of live microorganisms which, applied to animal or man, affect beneficially the host by improving the properties of the indigenous microflora' (Havenaar and

Huis in 't Veld, 1992). This implies that the term 'probiotic' is restricted to products which:

- contain live micro-organisms, e.g. as freeze-dried cells or in a fermented product;
- improve the health status of man or animals (which can include growth promotion of animals);
- can have their effect in the mouth or gastrointestinal tract (e.g. applied in food or capsules), but also in the upper respiratory tract (aerosal) or in the urogenital tract (local application).

An International Study Group on New Antimicrobial Strategies (ISGNAS) recognized that the development of microbial resistance to antibiotics is becoming a serious worldwide problem. This group is discussing the feasibility of developing rational alternatives to the use of antibiotics in the control of potentially pathogenic micro-organisms (Araneo *et al.*, 1996). It is worthwhile discussing the possibilities of probiotics as an alternative for therapeutic and growth-promoting antibiotics currently being used for poultry and pigs.

8.4.1 Probiotics used for pigs

For farm animals such as pigs, the most important reason for using a probiotic product is the claim (Gedek, 1991) that it influences the indigenous microflora resulting in:

- growth promotion of the animals in combination with a better feed conversion;
- health control, especially control of intestinal disturbances in young animals by stimulation of the development of the indigenous microflora and the colonization resistance in the gut;
- stimulation of the (non-)specific immune response;
- predigestion of antinutritional factors (ANFs) such as phytic acid, glucosinolates, trypsin inhibitors, lectins and non-starch polysaccharides (dietary fibre).

As stated in the introduction, the indigenous microflora in the gut inhibits the colonization of non-indigenous micro-organisms. Nevertheless, in some cases potentially pathogenic micro-organisms are able to penetrate and/or colonize these body surfaces, due to a massive attack of the pathogens or to a temporarily reduced colonization resistance. In different studies beneficial micro-organisms are investigated in order to demonstrate the improvement of the colonization resistance in pigs and other animals (reviewed in several papers, e.g. Fuller, 1986, 1989; Fox, 1988; Sissons, 1989; Havenaar and Huis in 't Veld, 1992) and/or the stimulation of the immune system, especially by lactobacilli (reviewed by Havenaar and Spanhaak, 1994).

Although the scientific evidence for the efficacy of probiotics in improving the health of pigs is rather poor, various probiotic products are on the market. The crucial questions now are: which specific microbial strains (probiotics) are responsible for one or more of these specific beneficial properties; how can one identify and select the most beneficial strains; and what is the best period to start with the probiotic?

The strategy for the selection and testing of probiotic strains has been published previously (Havenaar, Ten Brink and Huis in 't Veld, 1992). For identifying the best period to start with probiotics we performed studies in newborn piglets to investigate the composition of the gut microflora associated with the mucosa during maturation of this micro-flora. A critical point was determined as the period immediately after weaning: the day after weaning the numbers of lactobacilli decreased dramatically (approx. 1000-fold), while the number of *E. coli* increased far above the numbers of lactobacilli (Fig. 8.1; Huis in 't Veld and Havenaar, 1993). Similar effects of weaning on the composition of the gut microflora were found by Methew *et al.* (1996) – a significant decrease in the numbers of lactobacilli (30–1000-fold) and a significant increase in the numbers of coliforms – while the percentage of haemolytic coliforms increased with the age of the pigs. This rapid shift in

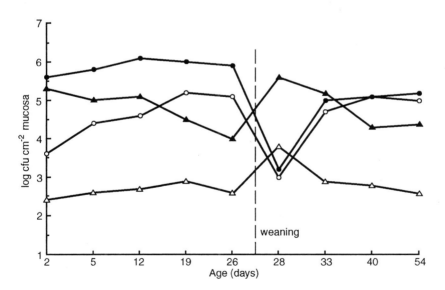

Fig. 8.1 Averge log numbers of *Lactobacillus* (○, ●) and *Escherichia coli* (△, ▲) per cm² jejunal (open markers) and ileal (closed markers) mucosa of piglets (*n* = 4) after birth and before and after weaning at 27 days of age (Huis in 't Veld and Havenaar, 1993).

micro-organisms after weaning, probably induced by stress and decreased uptake of feed with a different composition, could lead to overgrowth of enteropathogenic *E. coli*, resulting in postweaning diarrhoea.

To prevent this change in microflora two groups of piglets were treated with two different combinations of lactobacilli, isolated from healthy pigs. Each combination contained three different *Lactobacillus* strains which were carefully selected, based on *in vitro* experiments. The results showed that the probiotics induced stable numbers of lacto-bacilli on the small intestinal mucosa surface before and after weaning, while the numbers of *E. coli* dropped after supplementation of the probiotic strains and stayed at a similar low level after weaning (less than 10^3 cfu cm^{-2}) (Fig. 8.2; Huis in 't Veld and Havenaar, 1993).

Reduced faecal numbers of *E. coli* by the feeding of viable *Lactobacillus lactis* concentrate to piglets (0–54 days of age) was reported by Muralidhara *et al.* (1977). In a recently published study by Tortuero *et al.* (1995), reduced numbers of coliforms were found in the ileal contents of piglets by feeding the animals *Lactobacillus casei* plus *Enterococcus faecium*. No significant effects of lactic acid bacteria administra-

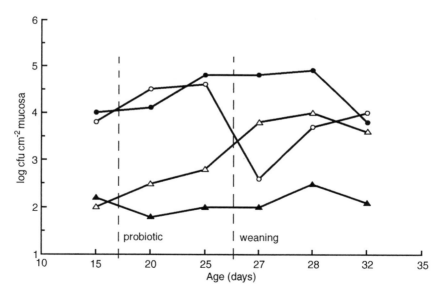

Fig. 8.2 Average log numbers of *Lactobacillus* (○, ●) and *Escherichia coli* (△, ▲) per cm^2 small intestinal mucosa of piglets ($n = 4$) before and after weaning in the control group (open markers) and a group receiving a probiotic containing $2–4 \times 10^6$ cfu g^{-1} feed of three different strains of lactobacilli (closed markers) from day 17 to 35 days of age (weaning at 26 days of age) (Huis in 't Veld and Havenaar, 1993).

tion were found on blood parameters, while the interleukin-2 concentration and the phagocytic activity in the ileal tissues showed an increase, although not significantly. Challenge experiments in pigs with these combinations of lactobacilli showed that the severity of disease manifestation (Fig. 8.3), induced by the infection with enteropathogenic *E. coli* K88 strains, as well as the intercurrent death of the animals, including euthanized piglets because of the seriousness of the disease (Fig. 8.4), were significantly lower in the probiotic group as compared with the control group (Havenaar, 1993). However, the numbers of *E. coli* in the intestinal contents as well as associated with the intestinal mucosa in the probiotic and control groups were not significantly different. Other researchers have demonstrated the reducing effect of feeding viable lactobacilli on the numbers of haemolytic coliform bacteria and also on the incidence of scouring in pigs (Muralidhara *et al.*, 1977; Zemanovic and Miljkovic, 1976). From these studies it can provisionally be concluded that:

- the composition of the intestinal microflora is influenced by external conditions such as weaning, resulting in a flora imbalance;
- orally administered selected *Lactobacillus* strains can result in a stable mature intestinal microflora, which is able to resist external influences such as weaning;

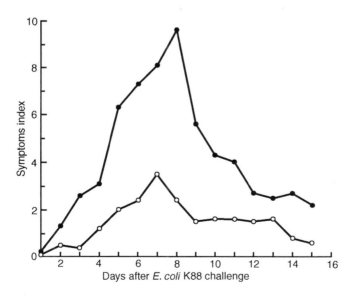

Fig. 8.3 Average severity of symptoms (expressed as an index) of piglets ($n = 20$ per group) after challenge with enteropathogenic *Escherichia coli* K88 in the control group (●) and a group receiving a probiotic containing $2-4 \times 10^6$ cfu g^{-1} feed of three different strains of lactobacilli (○) (Havenaar, 1993).

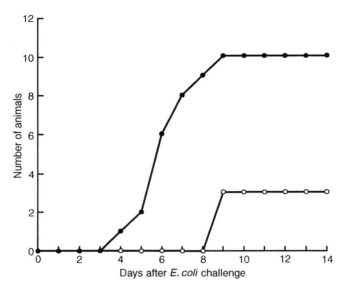

Fig. 8.4 Cumulative intercurrent death of piglets after challenge with enteropathogenic *Escherichia coli* K88 in the control group (●) and a group receiving a probiotic containing $2–4 \times 10^6$ cfu g^{-1} feed of three different strains of lactobacilli (○) (Havenaar, 1993).

- with selected lactobacilli a probiotic product can be made with a proven activity against enteropathogens.

The interaction between a *Lactobacillus* spp. and enteropathogenic *E. coli* K88 in gnotobiotic pigs was studied by Bomba *et al.* (1996). They found no prevention of the adherence of *E. coli* to the jejunal and ileal mucosa by the *Lactobacillus* spp. and in both groups of animals the pronounced clinical signs and intercurrent death were similar. The mechanism of the possible health-promoting effect of these lactobacilli is not yet explained. Further studies, such as field trials, should confirm these findings and elucidate the mechanism(s) of bacterial strains with proven activity against enteropathogens.

In vivo studies in ileal-fistulated pigs with the commercial products LBC-ME (Cernitin, Switzerland), containing *Enterococcus* (= *Streptococcus*) *faecium*, strain SF68 and Toyocerin (ToyoJozo, Japan), containing spores of *Bacillus toyoi*, showed that the bacteria in LBC-ME and Toyocerin survived passage through the stomach and small intestine: approximately 10^7 *Ent. faecium* (Fig. 8.5) and $10^4–10^5$ *B. toyoi* (Fig. 8.6) per gram chymus were found in the ileum (Havanaar and Huis in 't Veld, 1993). Based on the numbers of *Ent. faecium* or *B. toyoi* in the feed and the degree of dilution by drinking water and secreted oral and

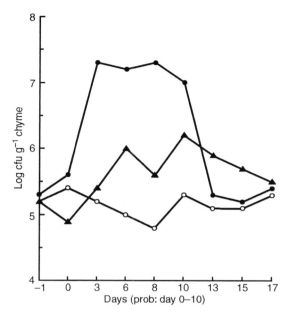

Fig. 8.5 Average log numbers of *Enterococcus faecium* (cfu g^{-1} ileal chyme) in ileal samples from fistulated piglets ($n = 4$) in a group feeding the probiotic LBC-ME (5×10^6 *E. faecium*/g feed) from day 0 to 10 (●) and a group feeding LBC-ME in combination with tylosin (50 ppm) from day 0 to 10 (▲) in comparison to a control group (○) without probiotic and tylosin (Havenaar, 1993).

gastrointestinal fluids, it can be suggested that *Ent. faecium* multiplies in the small intestine by a factor of 50–100, while *B. toyoi* showed no increase or decrease. After the probiotic was left out from the feed, the numbers of both bacterial species dropped to the basic levels, which indicates that both probiotic strains did not colonize the small intestine. Samples from the mucosa of the small intestine showed that *Ent. faecium* was associated with the mucosa side, in contrast to *B. toyoi*, which could not be isolated from the mucosa, but only from the intestinal contents. However, histological slides (light microscopy) did not show adherence of *Ent. faecium* to the brush borders. Therefore, the bacteria were probably embedded in the mucus layer. If tylosin was added to the feed in combination with the probiotic bacteria, the numbers of *Ent. faecium* did not increase significantly as compared with the basic levels and the numbers in the control group (Fig. 8.5). Apparently the growth of *Ent. faecium* in the intestine was inhibited by tylosin. The numbers of *B. toyoi* in the ileum were not influenced by tylosin (Fig. 8.6).

A challenge experiment in piglets showed that the infection with

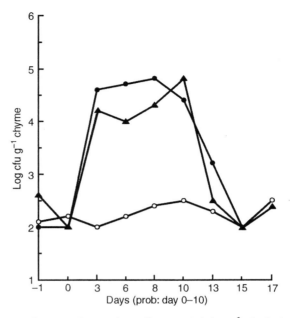

Fig. 8.6 Average log numbers of *Bacillus toyoi* (cfu g^{-1} ileal chyme) in ileal samples from fistulated piglets ($n = 4$) in a group feeding the probiotic Toyocerin (1×10^5 *B. toyoi* g^{-1} feed) from day 0 to 10 (●) and a group feeding Toyocerin in combination with tylosin (50 ppm) from day 0 to 10 (▲) in comparison to a control group (○) without probiotic and tylosin (Havenaar, 1993).

enteropathogenic *E. coli* resulted in intestinal disturbances, irrespective of the ingested *Ent. faecium* or *B. toyoi* administered in the feed. The increase in the average severity index of the symptoms up to day 9–15 and the decrease thereafter followed a similar pattern in the infected control group and in both probiotic groups (Fig. 8.7). The numbers of haemolytic *E. coli* in the contents (duodenum: 1–6 × 10^3 ml^{-1}; ileum: 2–8 × 10^5 ml^{-1}) and on the ileum mucosa (5–9 × 10$^{4-2}$) of the small intestine were not significantly different between the challenged control and probiotic groups (Havenaar and Huis in 't Veld, 1993).

Field trials with *Ent. faecium* strain 68 (Cernivet-LBC) resulted in reduced numbers of haemolytic *E. coli* (0141 K85ab) from herds suffering from coli-enterotoxaemia, but not in a significant effect on the clinical symptoms of the disease (Deprez *et al.*, 1989). The oral administration of *Ent. faecium* to pregnant sows (twice daily) resulted in lower incidences of scouring in the offspring as compared to the control group without probiotic (Maeng, Kim and Shin, 1989). The same strain of *Ent. faecium* also reduced the numbers of pathogenic *E. coli* and the

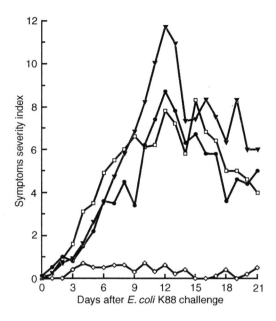

Fig. 8.7 Average severity of symptoms (expressed as an index) of piglets ($n = 15$ per group) after challenge with enteropathogenic *Escherichia coli* K88 in the control group (□), in two groups receiving the probiotic LBC-ME (●) or Toyocerin (▼) in comparison to a non-*E. coli*-infected control group (◇) (Havenaar and Huis in 't Veld, 1993).

severity of the induced disease in gnotobiotic piglets (Underdahl *et al.*, 1982). However, in this study the animals were only associated with *Ent. faecium* before the challenge with enteropathogenic *E. coli*, while the control group was not associated with any micro-organism before challenge. Germfree animals have a strongly reduced development of the intestinal tissue and the immune system. The daily administration of another strain of *Ent. faecium* (BIO-4R) to piglets resulted in a significant 20-fold reduction in the numbers of coliforms in faecal samples (Ozawa *et al.*, 1983).

Claims are made that probiotics may optimize the microbial ecology in the gastrointestinal tract and consequently promote the general health of the host and/or the resistance against enteropathogens. The majority of the commercial probiotics make claims such as growth promotion, without unequivocal scientific data and information about the underlying mechanism(s). Selection criteria can only be developed if they are based on scientific evidence which provides the explanation for the expected probiotic effect. Besides mechanisms, appropriate *in vitro* and animal models must be available for screening and validation, respectively. Finally, sensitive and reproducible (microbiological) methods are necessary for identification of the specific probiotic strain(s) in a complex microflora and to quantify the expected effects. Only in this way can the credibility of the probiotic concept be improved.

8.4.2   Use in poultry

Table 8.1 gives the results of trials with commercially available probiotic products after continuous (0–4 weeks) administration to broilers through the feed or drinking water. This period was chosen because colonization of potentially pathogenic micro-organisms occurs during the early days of life of the broilers and if positive effects of this intervention strategy tool are expected, these should become evident during this period. The group of Enterobacteriaceae was chosen as an indicator for the effects on *Salmonella* bacteria.

From these experiments it is concluded that the administration of probiotic products, at the given dosage during the 4-week trial period, did not influence body weight. The initial numbers of lactobacilli and other organisms present in the probiotic products seem not to exert any effect on the level of contamination in the caecal contents. Additional pH measurements in several parts of the gastrointestinal tract showed a large variation and no effect related to the administration of probiotic products could be observed. These trials led to the conclusion that probiotic compounds used under these conditions have no effect and are no tool in the intervention strategy against *Salmonella* colonization (see also Carazzoni *et al.*, 1993).

The health of birds with a stable gastrointestinal microflora is maintained by using antibiotics in preventive dosages. Low dosages of antibiotics have under certain conditions shown to be effective as growth promotors, but these low dosages have negative effects on lactobacilli and other lactic acid-producing bacteria. Lactobacilli constitute the largest part of aerobic microflora in the gastrointestinal tract and are very susceptible to antibiotics. To decrease the use of antibiotics, which may result in bacterial resistance and in formation of residues in organs and tissues of treated birds, the use of probiotics should be considered. There are some differences in the time-span in which action occurs. (Antibiotics are active in the short term, whereas the effects of probiotics may last days or weeks.) The name 'probiotics' is confusing and some years ago the suggestion, by a panel of European scientists, was made to call these products Ecological Health Control Products.

Lactobacillus reuteri could be the sole organism able to exert colonization resistance. *L. reuteri* is the only *Lactobacillus* species common to the gastrointestinal tract of humans and animals. *L. reuteri* produces and

Table 8.1 Commercially available probiotics and their effects on (a) body-weight and (b) Enterobacteriaceae and lactobacilli after a 4-week trial

(a)

Product	Micro-organism	Dosage (kg t^{-1})	Average weight (kg)	CV (%)
Fralac	*Streptococcus faecium* SF68	0.080	1.1097	10.1
FraSacc	*Saccharomyces cerevisiae*	0.400	1.1095	8.8
Gist Brocades	Unknown yeast	2.000	1.0577	13.0
BioPlus	*Bacillus subtilis*	0.400	1.1378	7.3
	Bacillus licheniformis	0.500	1.1092	10.7
Toyocerin	*Bacillus toyoi*	0.500	1.0698	7.2
KemPro	*Saccharomyces cerevisiae*	1.000	1.1210	11.8
LactoSacc	*Lactobacillus, Streptococcus, yeast, enzymes*	1.000	1.1202	10.4
BioSavor	*Lactobacillus acidophilus Lactobacillus plantarum*	0.005	1.0938	13.3
AllLac	*Lactobacillus, Streptococcus*	0.025	1.2021	9.6

(b)

Product name	Enterobacteriaceae contents after week		Lactobacilli contents after week	
	1	4	1	4
Control	<9.0	7.9	9.0	5.0
Fralac	>9.0	7.7	5.9	5.2
FraSacc	>9.0	7.9	7.0	5.2
Gist Brocades	>9.0	7.7	6.5	6.1
BioPlus	>9.0	7.9	6.8	5.3
Toyocerin	>9.0	6.9	7.0	<5.0
KemPro	>9.0	7.3	8.5	5.2
LactoSacc	>9.0	8.9	7.8	<5.0
BioSavor	>9.0	8.9	7.8	<5.0
AllLac	>9.0	6.9	6.9	<5.0

secretes (in the presence of glycerol) an intermediary metabolite which has antimicrobial activity. This metabolite, termed reuterin, was shown to have broad-spectrum antimicrobial activity against bacteria such as *Salmonella*, *E. coli* and *Campylobacter*. *L. reuteri* is the only bacterial species known to secrete this metabolite into the surrounding microenvironment in amounts sufficient to impart antimicrobial activity. *L. reuteri* shows a high degree of resistance to gastric acid and bile salts. This enables the organism to survive passage through the stomach and thus colonize the gastrointestinal tract in humans and animals.

Animal models are used to demonstrate the health benefits, for example protective effects, of *L. reuteri* to the host. *L. reuteri* prophylaxis has been shown to moderate stress-associated diseases. Young chickens were protected against deaths associated with exposure to a *Salmonella typhimurium* challenge. In control birds approximately 40% died after challenge, in *L. reuteri* treated birds this percentage was about 5%. Similar effects were observed with young turkeys. Also with piglets *L. reuteri* has shown very beneficial effects as treatment against diarrhoea induced by *Cryptosporidium parvum*. The pig is often used as an animal model for translation of results to humans as its gastrointestinal tract is very similar to the human gastrointestinal tract. Therefore the positive results of a trial conducted in Finland focusing on diarrhoeal disease in young children are not surprising.

8.5   USE OF COMPETITIVE EXCLUSION MICROFLORAS IN POULTRY

Competitive exclusion (CE) is the name for the treatment of day-old chicks with a microflora resulting in a colonization resistance towards potentially pathogenic micro-organisms with special reference to *Salmonella* colonization of the gut. In contrast to the continuous administration of probiotics, the competitive exclusion treatment is given only once, although a second or third dosage at a later age is considered to be beneficial (Bilgili, 1995; Burns, 1995).

Whereas the composition of a probiotic product is defined – products consist of pure cultures of one or more micro-organisms – the composition of the CE microfloras which have proven to be effective is not completely known. British and Canadian researchers have studied the effect of CE microflora consisting of 50 different pure cultures on the numbers and ratio of their presence in the gastrointestinal tract of broilers, but the results were less promising than those obtained after the administration, through the drinking water, of undefined microfloras. Examples are the products produced by Finnish, Dutch, American and British researchers. These undefined microfloras are homogenates from caeca and caecal contents and contain a large number of known aerobic micro-organisms and a number of unknown, mainly anaerobic, micro-organisms. The initial observation that one specific segmented micro-organism would be responsible for the colonization resistance against *Salmonella* could not be confirmed.

The safety of these CE products is guaranteed, and during the past 15 years of research and practical application no adverse effects on bird performance and zootechnical data have been observed following their use. The use of CE microfloras, several brands of which are commercially available, as part of an intervention strategy against *Salmonella* contamination is accepted. Although, as the results in Tables 8.2–8.4 make clear, the use of these products will not yet lead to a

Table 8.2 Effect of competitive exclusion (CE) treatment under practical conditions on *Salmonella* contamination of caeca of broilers (Goren *et al.*, 1991)

Treatment	Number of samples	Number positive
Non-CE treated	14 099	486 (3.5%)
CE treated	14 400	134 (0.9%)

Total 284 flocks, 8 million broilers

Table 8.3 Effect of competitive exclusion (CE) treatment under practical conditions on *Campylobacter* contamination of broiler flocks (Mulder and Bolder, 1991)

Treatment	Number of flocks	Number positive
Non-CE treated	29	18 (62%)
CE treated	29	12 (41%)

Total 200 000 broilers

Table 8.4 *Salmonella* and *Campylobacter* incidence in flocks (%) after treatment with Broilact under practical conditions (Bolder *et al.*, 1995)

	Salmonella		*Campylobacter*
	Cloacal swab samples at 4 weeks of age	Caecal contents at slaughter	Caecal contents at slaughter
Control	22.8	23.7	44.8
Broilact	12.3	19.8	32.4
Total	17.0	21.7	38.4

zero level of *Salmonella* contamination in the consumer-ready products, their application is seen as an important development in the quest for final elimination of *Salmonella* from the chicken gut.

Undefined microfloras, although effective and proven free of pathogens, have the disadvantage that legislators cannot accept their implementation easily. However, the main problem in the use of defined microfloras is how to sustain the viability and effectiveness over a long period of time. These problems seem to be solved with undefined products.

The use of CE microfloras against *Salmonella* contamination is proven to be effective. The administration of the flora in the hatchery, however, causes some opposition, as it is very illogical to spray a large number of micro-organisms (albeit harmless) into an environment which has been kept more or less sterile to prevent contamination. Therefore, new ways of application are under study, one of them being the *in-ovo* injection.

However, application of CE microfloras should also change farmers' and veterinarians' attitudes to the use of antibiotics. Several antibiotics are known to undermine the effectiveness of CE microfloras. The use of several others has an indifferent effect on *Salmonella* colonization. The mechanisms of colonization or protection from colonization are not yet answered. However, the application of CE microfloras results in a considerable decrease of the contamination rate of the live bird. To maintain the beneficial effect of CE treatment to the consumer-ready end product, additional improvements in hygiene, in transportation and slaughtering are also necessary.

8.6 SUMMARY

Healthy animals possess a natural resistance against colonization or infection by potentially pathogenic micro-organisms. The host–gastrointestinal flora interaction is responsible for this phenomenon. Over the years the conditions of animal production have changed, also influencing these natural resistance characteristics. Breeding and husbandry conditions changed; more efficient and therefore more productive animals have been used and the use of antimicrobial compounds has increased and, as a result of this, production methods have become increasingly more intense.

These changes led to stress conditions in the animals as well as deficiencies in the composition of the gastrointestinal microflora, resulting in less natural resistance of the animals against contamination or infection with pathogens. Especially in the case of poultry, there may be a delayed formation of the gastrointestinal microflora because of the lack of contact with maternal and/or environmental micro-organisms.

Meat and meat (including poultry) products can be contaminated with potentially pathogenic micro-organisms such as *Salmonella* and *Campylobacter*. The role of meat and meat products in human foodborne disease also is proven and therefore, as a very important aspect of food safety, the industry is urged to decrease the level of contamination within the near future to zero or at least to acceptable levels. Several intervention strategies have been proposed starting at the breeding and farm level through to the consumer-ready product. Part of these intervention strategies may be the use of probiotic and competitive exclusion microfloras. These products have been proposed for prophylactic and curative use. This chapter discusses the various factors influencing the colonization of micro-organisms and gives some data on the practical use of products as part of an intervention strategy against *Salmonella* contamination in pig and poultry production.

8.7 REFERENCES

Anon (1996) *Lactobacillus reuteri* and the BRA concept. Brochure Biogaia Biologics 1996-02-25

Areneo, B.A., Cebra, J.J., Beuth, J. *et al.* (1996) Problems and priorities for controlling opportunistic pathogens with new antimicrobial strategies: an overview of current literature. *Zentralblatt Bakteriologie – Int. J. Med. Microbiol. Virology Parasitol. Inf. Dis.*, **283** (4), 431–65.

Bilgili, S.F. (1995) Competitive exclusion: the basics. *Broiler Industry*, September, 18.

Bolder, N.M., P.F.G. Vereijken., Putirulan, F.F. and Mulder, R.W.A.W. (1995) The effect of competitive exclusion on the Salmonella contamination of broilers (a field study). *Proceedings XII European Symposium on the Quality of Poultry Meat*, Zaragoza, Spain, pp. 89–98.

Bomba, A., Nemcova, R., Kastel, R. *et al.* (1996) Interactions of *Lactobacillus* spp and enteropathogenic *Escherichia coli* under *in vitro* and *in vivo* conditions. *Vet. Medicina*, **41** (5), 155–8.

Burns (1995) Commercial competitive exclusion bacterial mix ready soon. *Feedstuffs*, **67**, no. 11.

Cavazzoni, Adami, Castrovilli, Succi (1993) A preliminary experimentation on broilers with a strain of *Bacillus coagulans* as probiotic. *Microbiologie-Aliments-Nutrition*, **11**, 457–62.

Chesson, A. (1994) Probiotics and other intestinal mediators, in *Principles of Pig Sciences* (eds D.J.A. Cole, J. Wiseman and M.A. Varley), Nottingham University Press, Loughborough, UK.

Collington, G.K., Parker, D.S. and Armstrong, D.G. (1990) The influence of inclusion of either an antibiotic or a probiotic the diet on the development of digestive enzyme activity in the pig. *Br. J. Nutrition*, **64**(1), 59–70.

Deprez, P., van den Branden, J., De Geest, J. *et al.* (1989) De invloed van *Streptococcus faecium* toediening op de excretie van *Escherichia coli* en het voorkomen van slingerziekte bij gespeende biggen. *Vlaams Diergeneesk. Tijdschr.*, **58**, 113–17.

Gedek, B. (1991) Probiotika in der Schweinehaltung. *DGS*, **43**, 1052–5.

Gillingham (1994) Probiotics: a concept for the future of poultry health management. *Canada Poultryman*, April, 22–4.

Goren, E., de Jong, W.A., Doornenbal, P., Bolder, N.M., Mulder, R.W.A.W. and Jansen, A. (1988) Reduction of *Salmonella* infection of broilers by spray application of intestinal microflora: a longitudinal study. *Vet. Quart.*, **10**, 249–54.

Fox, S.M. (1988) Probiotics: intestinal inoculants for production animals. *Vet. Med.*, **9**, 806–10, 823–30.

Fuller, R. (1986) Probiotics. *J. Appl. Bacteriol.*, **63**(Suppl.), 1S–7S.

Fuller, R. (1989) Probiotics in man and animals. *J. Appl. Bacteriol.*, **66**, 365–78.

Havenaar, R. (1993) Bacterial strains with proven activity against enteropathogens, in *Interruption of Bacterial Cycles in Animal Production: Related to Veter-*

inary Public Health (ed. B.A.P. Urlings), Workshop 8–10 December 1993, Utrecht, The Netherlands. ADDIX, Wijk bij Duurstede, The Netherlands, pp. 97–109.

Havenaar, R. and Huis in 't Veld, J.H.J. (1992) Probiotics; a general view, in *The Lactic Acid Bacteria, Vol. I. The Lactic Acid Bacteria in Health and Disease* (ed. B.J.B. Wood), Elsevier Applied Science, Barking, pp. 151–70.

Havenaar.R. and Huis in 't Veld, J.H.J. (1993) *In vitro* and *in vivo* experiments with two commercial probiotic products containing *Enterococcus faecium* and *Bacillus toyoi*, in *Probiotics and Pathogenicity* (eds J.F. Jensen, M.H. Hinton and R.W.A.W. Mulder), COVP-DLO Het Spelderholt, Beekbergen, The Netherlands, pp. 54–62.

Havenaar, R. and Spanhaak, S. (1994) Probiotics from an immunological point of view. *Curr. Opin. Biotechnol.*, **5**, 320–5.

Havenaar, R., Ten Brink, B. and Huis in 't Veld, J.H.J. (1992) Selection of strains for probiotic use, in *Probiotics: The Scientific Basis* (ed. R. Fuller), Chapman & Hall, London, pp. 209–24.

Huis in 't Veld, J.H.J. and Havenaar, R. (1993) Selection criteria for microorganisms for probiotic use, in *Probiotics and Pathogenicity* (eds J.F. Jensen, M.H. Hinton and R.W.A.W. Mulder), COVP-DLO Het Spelderholt, Beekbergen, The Netherlands, pp. 11–19.

Lee, Y.K. and Salminen, S (1995) The coming age of probiotics. *Trends Food Sci. Technol.*, **6**, 241–5.

Maeng, W.J., Kin, C.W. and Shin, H.T. (1989) Effect of feeding lactic acid bacteria concentrate (LBC, *Streptococcus faecium* Cernelle 68) on the growth rate and prevention of scouring in piglet. *Korean J. Anim. Sci.*, **31**, 318–23.

Mathew, A.G., Franklin, M.A., Upchurch, W.G. and Chattin, S.E. (1996). Influence of weaning age on ileal microflora and fermentation acids in young pigs. *Nutrition Res.*, **16**(5), 817–27.

Metchnikoff, E. (1907) *The Prolongation of Life*. E.P. Putman's Sons, New York.

Miller, B.G., James, P.S., Smith, M.W. and Bourne, F.J. (1986) Effect of weaning on the capacity of pig intestinal villi to absorb nutrients. *J. Agric. Sci., Cambridge*, **107**, 579–89.

Mulder, R.W.A.W. and Bolder, N.M. (1991) Reduction of *Campylobacter* infection in broilers by competitive exclusion treatment by day-old chicks. A field study, in *Colonization Control of Human Bacterial Enteropathogens in Poultry* (ed. L.C. Blankenship), Academic Press, London, pp. 359–63.

Muralidhara, J.S., Sheggeby, G.C., Elliker, P.R. *et al.* (1977) Effect of feeding lactobacilli on the coliform and lactobacillus flora of intestinal tissue and faeces from piglets. *J. Food Prot.*, **40**, 480–5.

Nouusiainen, J. and Suoni, K. (1991) Comparative observations on selected probiotics and olaquindox used as feed additive for piglets around weaning. 2. Effect of villus length and crypt depth in the jejunum, ilium, cecum and colon. *J. Anim. Physiol. Anim. Nutr.*, **66**, 224–30.

Ozawa, K., Yabu-Uchi, K., Yamanaka, K. *et al.* (1983) Effect of *Streptococcus faecalis* BIO-4R on intestinal flora of weanling piglets and calves. *Appl. Environ. Microbiol.*, **45**, 1513–18.

Salminen, S., Isolauri, E. and Salminen, E. (1996) Clinical use of probiotics for stabilizing the gut mucosal barrier: successful strains and future challenges. *Anthonie van Leeuwenhoek*, **70**, 347–58.

Sissons, J.W. (1989) Potential of probiotic organisms to prevent diarrhoea and promote digestion in farm animals – a review. *J. Sci. Food Agric.*, **49**, 1–13.

Tortuero, F., Rioperez, J., Fernandez, E. and Rodriguez, M.L. (1995) Response of piglets to oral administration of lactic acid bacteria. *J. Food Protection.*, **58**, 1369–74.

Underdahl, N.R., Torres-Medina, A. and Doster, A.R. (1982) Effect of *Streptococcus faecuim* C-68 in control of *Escherichia coli*-induced diarrhea in gnotobiotic pigs. *Am. J. Vet. Res.*, **43**, 2227–32.

White, D.D. and Savage, D.C. (1987) Lactobacilli as effectors of host functions: no influence on the activities of enzymes in enterocytes of mice. *Appl. Env. Microbiol.*, **53**, 325–30.

Wick, M.J., Madara, J.L., Fields, B.N. and Normark, S.J. (1991) Molecular cross talk between epithelial cells and pathogenic microorganisms. *Cell*, **67**(4), 651–9.

Zemanovic, M. and Miljkovic, V. (1976) Application of *Lactobacillus acidophilus* in the prophylaxis and therapy of diarrhoea in piglets. *Proc. 4th Int. Pig Vet. Soc. Congr., Ames, Iowa*, p. 121.

Index